DESIGN OF HIGH-SPEED
COMMUNICATION CIRCUITS

SELECTED TOPICS IN ELECTRONICS AND SYSTEMS

Editor-in-Chief: **M. S. Shur**

Selected Topics in Electronics and Systems – Vol. 38

DESIGN OF HIGH-SPEED COMMUNICATION CIRCUITS

Editor

Ramesh Harjani
University of Minnesota, USA

World Scientific

NEW JERSEY • LONDON • SINGAPORE • BEIJING • SHANGHAI • HONG KONG • TAIPEI • CHENNAI

Published by

World Scientific Publishing Co. Pte. Ltd.

5 Toh Tuck Link, Singapore 596224

USA office: 27 Warren Street, Suite 401-402, Hackensack, NJ 07601

UK office: 57 Shelton Street, Covent Garden, London WC2H 9HE

British Library Cataloguing-in-Publication Data
A catalogue record for this book is available from the British Library.

DESIGN OF HIGH-SPEED COMMUNICATION CIRCUITS

ISBN 981-256-590-6

Editor: Tjan Kwang Wei

Printed in Singapore by Mainland Press

Preface

Welcome to the special issue of the International Journal of High Speed Electronics and Systems on "High-Speed Mixed-Signal Integrated Circuits". Silicon, in particular MOS silicon, has rapidly become the *de facto* technology for mixed-signal integrated circuit design due to the high levels of integration possible as device geometries have shrunk to nanometer scales. The reduction in feature size has meant that the number of transistor and clock speeds have increased significantly. In fact, current day microprocessors contain hundreds of millions of transistor and operate at multiple giga Hertz. Further, this reduction in feature size has also had a significant impact of mixed-signal circuits. Due to the higher levels of integration the majority of ASICs have some analog component on them. Additionally, it has now become nearly mandatory to integrate both analog and digital circuits on the same substrate due to cost and power constraints. The eight manuscripts in this special edition focus on the some of the newer problems and opportunities offered by the small device geometries and the high levels of integration that is now possible.

This volume opens with an introduction to the analog issues surrounding accuracy concerns in nanometer CMOS by M. Flynn, S. Park and C.C. Lee. This paper reviews some of the causes of and trends in MOS transistor mismatch, and assess the implications for analog circuit design in the nanometer régime. Device matching improves as CMOS technology evolves but reductions in power supply voltages associated with the smaller geometries complicates design. Unlike digital circuits, in analog circuits lower power supplies result in higher power consumption due to thermal noise limits. However, new circuit techniques based on analog circuit redundancy can avoid the accuracy-power constraints related to device mismatch. To illustrate such techniques this paper focuses on data converters.

One of the consequences of the higher integration levels is the fact that analog and RF circuits now have to live on the same substrate with vast amounts of noisy digital circuits. R. Gharpurey and S. Naraghi introduce us to the issues and design techniques required to survive in the presence of such self-induced noise. Electromagnetic coupling of noisy circuits in close physical proximity with sensitive circuits can result in harmful interference and severely degrade performance. In this paper, the authors discuss the evolution of techniques for modeling and analyzing the sources of noise generation and interference and provide techniques for the extraction of electrical models. They introduce the concepts of functional modeling of circuit functions and present a model for an integrated flash ADC.

Noise-shaped oversampling has been used for a number of years as a mechanism to extract high resolution at the expense of higher sampling frequencies in data converters. A. Gharbiya, T.C. Caldwell and D.A. Johns provide an overview of oversampled ADCs that are particularly well suited for high-speed operation. The authors first discuss various discrete-time architectures and describe their performance limitations. The paper then discusses how time-interleaving can be used to improve signal bandwidth. Continuous-time operation for oversampled converter has often been proposed as a mechanism to improve signal bandwidth. The paper then describes design tradeoffs and issues associated with continuous-time operation and conclude with a summary of recent state of the art high-speed converters.

Integrated LC voltage-controlled oscillators are critical building blocks in both wireless and wireline communication systems. They often set the performance limits of such systems. B. Jung and R. Harjani describe design techniques that use a capacitively degenerated negative resistance cell for high-frequency high-performance VCOs. The authors show that a negative resistance cell using capacitive degeneration has a higher maximum oscillation frequency and a smaller equivalent shunt capacitance when compared to other widely used topologies. The authors then describe enhancements to the basic capacitive degeneration topology. These are followed by test results for a CMOS and a BiCMOS test vehicle that validates the efficacy of the capacitive degeneration technique.

Frequency synthesizers utilize VCOs as part of a phase-locked loop and are a key building block of fully-integrated wireless communications systems. S.T. Moon, A.Y. Valero-Lopez and E. Sanchez-Sinencio provide a tutorial for fully integrated frequency synthesizers. It is critical to fully appreciate both circuit and system-level design issues while designing such frequency synthesizers. In this tutorial article, the authors describe general implementation issues and recent developments of frequency synthesizer design. The authors provide simplified and intuitive explanations that help both initial design and consequential troubleshooting when problems arise.

In recent years one of the consequences of the small device geometries is the possibility of designing fully integrated CMOS wireless transceivers. D. Allstot, S. Aniruddhan, M. Chu, J. Paramesh and S. Shekhar provide a summary of some of the recent advances and design trends in such transceivers. They describe a number of traditional and modern transceiver architectures. This discussion is followed by details for a number of the building blocks including low-noise amplifiers, mixers, and voltage-controlled oscillators. The authors then go on to describe low phase noise circuit designs that are suitable for quadrature carrier generation.

As clock speeds for microprocessors and other computer devices have increased the I/O has rapidly become the bottleneck for high-performance computing. P.K. Hanumolu, G.Wei and U. Moon describe equalizers that are used in high-speed serial links. Equalization techniques are often used to mitigate ISI in communication links that result from finite bandwidth effects. In this article, the authors describe receive and transmit equalizers in both digital and analog form. Nonlinear equalization is often used to circumvent some of the limitations of linear equalization techniques. The authors describe DFE loop latency issues and conclude with adaptive algorithms and techniques.

As clock speeds and chip-to-chip and backplane operating speeds have increased so to has the need for high-speed disk I/Os. C.P. Yue, J. Park, R. Sun, L.R. Carley, and F. O'Mahony describe a high-speed parallel interface for disk drives. The authors' present low-power circuit techniques suitable for high-speed digital parallel interfaces each operating at over 10 Gbps between the channel IC and the magnetic read head. The design utilizes a crosstalk cancellation technique using a novel data encoding schemes and multi-level signal encoding. The authors go on to describe a number of the key circuit blocks including the receive sampler, the phase interpolator, and the transmitter output driver including measurement results from a prototype design.

Ramesh Harjani
Minneapolis, MN
2005

CONTENTS

International Journal of High Speed Electronics and Systems
Vol. 15, No. 2 (2005) 255–275
© World Scientific Publishing Company

ACHIEVING ANALOG ACCURACY IN NANOMETER CMOS

MICHAEL P. FLYNN, SUNGHYUN PARK, AND CHUN C. LEE

Electrical Engineering & Computer Science, University of Michigan, 1301 Beal Avenue,
Ann Arbor, Michigan 48109-2122, USA
mpflynn@eecs.umich.edu

This paper reviews causes of and trends in MOS transistor mismatch, and assesses the implications for analog circuit design in the nanometer régime. The current understanding of MOS transistor mismatch is reviewed. In most cases, transistor mismatch is dominated by threshold voltage mismatch. Although, there is strong evidence that V_T matching is improving as CMOS technology evolves, these improvements are countered by reductions in power supply that also accompany process scaling. In fact, the power consumption of analog circuits based on current design styles will increase with scaling to finer processes. It has long been known that thermal noise causes the power consumption of analog circuits to increase with scaling. However, unlike the case with thermal noise, new circuit techniques can break the accuracy-power constraints related to mismatch. These techniques are based on analog circuit redundancy, and take advantage of the tremendous transistor density offered by nanometer CMOS. This paper is primarily concerned with comparators, and in particular, with the use of comparators in flash ADCs; however, the analysis is also applicable to other circuits and applications.

Keywords: Analog circuits, CMOS, mismatch, offset, flash, ADC, offset cancellation.

1. Introduction

The performance of analog circuits depends on both the absolute value of component parameters, such as resistance, capacitance, and transconductance, and the matching of the parameter values for different components. Absolute parameter values may vary by 20% or more from die to die, and in addition, these parameters are sensitive to changes in temperature and supply voltage. Since the matching of component parameters on an integrated circuit is far better than the component accuracy, circuit designers have created analog signal processing schemes that rely on component matching rather than absolute values. As an example, switched capacitor circuits rely on capacitor ratios, rather than absolute values of capacitance.

The matching of MOS transistors is critical to many CMOS analog signal processing schemes. From a transistor matching point of view, analog circuit building blocks fall into two broad categories of operation; circuits that operate on voltage differences and circuits that generate matched or ratioed currents.[1] The latter category includes current mirrors, and current DACs, while the former includes voltage comparators and opamps. In this paper, we concentrate on voltage input circuits, dealing specifically with comparators and especially to their application in analog-to-digital conversion. Comparators are an integral part of analog-to-digital converters, and in fact, are required in most analog-digital interfaces, for example, in clock-data recovery circuits. However, our analysis and conclusions are also relevant to other types of circuits.

In this paper, we review the causes and some trends in transistor mismatch and assess the implications for analog circuit design. We begin by reviewing the current understanding of MOS transistor mismatch. In most cases, transistor mismatch is dominated by threshold voltage mismatch. Although, there is strong evidence that V_T matching is improving as CMOS fabrication evolves, these improvements are countered by reductions in power supply that also accompany process scaling. In fact, we show that the power consumption of analog circuits, based on current design styles, will increase with scaling to finer processes. It has long been known that fundamental limits associated with thermal noise will cause power consumption to increase with scaling. However, unlike the case with thermal noise, new circuit techniques can break the accuracy-power constraints related to mismatch. New techniques, such as analog redundancy, take advantage of the tremendous transistor density offered by nanometer CMOS.

2. Transistor Mismatch

Over the past two decades much work had been done on modeling and explaining MOS transistor mismatch. Early efforts were geared towards analog design, but recently, mismatch has also become a concern for digital designers. Since MOS transistors in analog circuits are most often operated in saturation, analysis of mismatch is usually done in this region. (Mismatch in subthreshold has also been examined.[2,3])

We deal here with random mismatch, but there are numerous sources of systematic offset which can be equally detrimental. (These systematic sources include stress effects due to packaging[4] and metal fill[5], temperature gradients,[6] and unmatched metal coverage.) We begin with the long channel drain current in saturation:

$$I_D = \frac{\beta}{2}\left(V_{GS} - V_T\right)^2, \text{ where } \beta = \mu C_{ox} W/L. \tag{1}$$

We can attribute the drain current mismatch in two identical transistors to differences in β (known as $\Delta\beta$) and V_T (called ΔV_T.). We also know that the threshold voltage is modified by the body effect when the source and bulk are not at the same potential (i.e. $V_{SB} \neq 0$):

$$V_T = V_{T0} + \gamma\left(\sqrt{V_{SB} + 2\Phi_F} - \sqrt{2\Phi_F}\right). \tag{2}$$

We separate mismatch in threshold voltage into mismatch ΔV_{T0} of the nominal threshold voltage V_{T0}, and mismatch $\Delta\gamma$ of the body effect parameter γ.

Following Pelgrom's presentation[7] (an earlier analysis of mismatch is also given[8]), mismatch in a device parameter P (e.g. β, V_{T0} or γ) can be separated into local and distance related effects. We model the variance in a parameter mismatch as:

$$\sigma^2(\Delta P) = \frac{A_P^2}{WL} + S_P^2 D_x^2, \tag{3}$$

where A_P and S_P are empirical parameters, W and L are the transistor dimensions, and D_x is the distance between the devices. The second component, describing the distance related variance of mismatch, can be attributed to gradual gradients in oxide thickness and doping across a die. Clearly, we would expect this mismatch contribution to become more pronounced as the distance D_x between devices increases. On the other hand, the local mismatch effect, described by the first term, is caused by random localized defects. The random defects that cause local mismatch tend to be averaged over the transistor gate area. It is therefore, reasonable to suppose that larger devices suffer less from mismatch, and as indicated in Eq. (3), the variance of the local term is inversely proportional to gate area WL. In this way, we can model mismatch in the three parameters V_{T0}, γ and β as:

$$\sigma^2(\Delta V_{TO}) = \frac{A_{VTO}^2}{WL} + S_{VTO}^2 D_x^2, \tag{4}$$

$$\sigma^2(\Delta\gamma) = \frac{A_\gamma^2}{WL} + S_\gamma^2 D_x^2, \tag{5}$$

$$\frac{\sigma^2(\Delta\beta)}{\beta^2} = \frac{A_\beta^2}{WL} + S_\beta^2 D_x^2. \tag{6}$$

A_{VT0}, A_γ and A_β are empirical parameters that must be measured for each CMOS process. It is worth noting the β term encompasses mismatches in mobility and oxide capacitance as well as mismatches in the gate dimensions W and L. For improved accuracy, the effective width (W_{eff}) and length (L_{eff}) are used in place of the drawn width and length W and L.[13]

Table 1. Historical and predicted mismatch information

Feature Size	V_{DD} (V)	A_{VT0} (mV.μm)	t_{ox} (nm)	C_{ox} (fF/ μm²)	$A_β$ (%.μm)	Source
2.5 μm	5.0	30	50	0.7	2.3	[9]
1.2 μm	5.0	21	24	1.4	1.8	[9]
0.7 μm	5.0	13	14	2.5	1.9	[9]
0.5 μm	5.0	11	10	3.5	1.8	[9]
0.35 μm	3.3	9	7	4.9	1.9	[9]
0.25 μm	2.5	6	5	6.9	1.85	[9]
0.18 μm	1.8	6	4	8.6	1.53	[9]
0.13 μm	1.2	5	2.3	15.0	--	ITRS '01
100 nm	1.0	6	2.1	16.4	--	[10]
90 nm	1.0	4	2	17.3	--	[10]
80 nm	0.95	4	1.8	19.2	--	[10]
70 nm	0.9	4	1.7	20.3	--	[10]
65 nm	0.85	3	1.3	26.6	--	[10]
57 nm	0.8	3	1.2	28.8	--	[10]
50 nm	0.75	3	1.2	28.8	--	[10]

Later, we consider the relative importance of the threshold voltage and current factor variance terms, but first we compare the importance of the local and distance terms. Table 1 shows mismatch parameters reported for several processes,[9] along with mismatch parameters predicted by the International Technology Roadmap for Semiconductors (ITRS) [10] for future processes. For convenience, these parameters are almost always formulated with transistor dimensions expressed in microns and not in meters. (For example, A_{VT0} has the units $mVμm$.). Except where the distance or gate areas are large, the local effect is much more important than the long distance effect. For a given transistor area, we can calculate the crossover distance at which the local and distance effects are equal. Since the local mismatch effects become smaller as we increase gate area, the cross over distance is smaller for larger devices. In Figure 1, we plot this breakpoint distance versus gate area for one CMOS process (using mismatch data for 2.5μm CMOS[7]). In most practical cases, we can ignore the distance effect, except in the case of large circuit blocks such as DAC or for global biasing.

Next, we consider the relative importance of the current factor and threshold voltage errors. We follow an approach similar to that of Pelgrom et al.[7] and Kinget et al.[11] Applying the MOS long-channel drain current equation; we combine the effects of current factor and threshold voltage mismatch.[7]

$$\frac{\sigma^2(\Delta I_d)}{I_d^2} = \frac{4\sigma^2(\Delta V_{T0})}{(V_{GS} - V_{T0})^2} + \frac{\sigma^2(\Delta\beta)}{\beta^2} \tag{7}$$

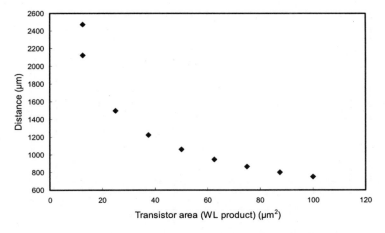

Fig. 1. Distance breakpoint for V_{T0} local versus long distance mismatch.

Substituting for the threshold voltage and current factor mismatches we get:

$$\frac{\sigma^2(\Delta I_d)}{I_d^2} = \frac{1}{WL}\left(A_\beta^2 + \frac{4A_{VT0}^2}{(V_{GS} - V_{T0})^2}\right). \tag{8}$$

We note from this equation that the relative importance of the two mismatch contributions depends on the overdrive voltage V_{GS}-V_{T0}. For low and moderate values of overdrive voltage the threshold voltage mismatch term dominates. Kinget[11] defines a breakpoint voltage of $(V_{GS}$-$V_T)_m$ at which both contributions are equal.

$$\left(V_{GS} - V_{T0}\right)_m = 2A_{VT0} / A_\beta. \tag{9}$$

Using the historical data, we plot the breakpoint voltage for a range of CMOS processes in Figure 2. In most cases we use the minimum value of overdrive, so that we make the best use of limited headroom and achieve the maximum transconductance for a given drain current. However, the overdrive voltage should also be greater than six times the thermal voltage (i.e. $6\ kT/q$ ~150 mV) to ensure a strongly inverted channel.

Assuming that threshold voltage mismatches dominate, we now review the causes of V_{T0} mismatch and explore trends. It is now well accepted that the most important factor in V_{T0} mismatch is variation in the number of donor and acceptor atoms ($N=Na+Nd$) in the depletion layer under the gate[12-14], and we can write:

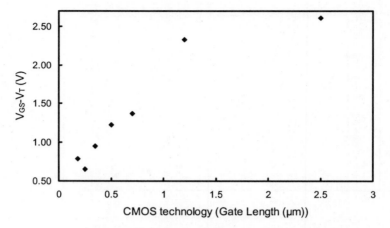

Fig. 2. Overdrive voltage at which β mismatch begins to dominate.

$$\sigma_{\Delta VT0} = \frac{A_{VT0}}{\sqrt{WL}} = \frac{qt_{ox}\sqrt{2Nt_{depl}}}{\varepsilon_0\varepsilon_{ox}\sqrt{WL}}.$$

(10)

Here t_{ox} is the gate oxide thickness, and t_{depl} is the depletion width. While fluctuations in N are important, several other effects are becoming significant, including interface states,[14] oxide roughness, and variations in W and L.[15] Nevertheless, there is strong experimental evidence that $\sigma_{\Delta VT}$ is proportional to t_{ox}, or equivalently, that A_{VT0} is inversely proportional to C_{ox}. A nominal trend of $A_{VT0} = 0.707$ mVμm per nanometer of t_{ox} has been reported[15]; however, it seems unlikely that this trend will continue. Before we conclude, we note that V_{T0} mismatch is largely independent of temperature.

3. Methods of Achieving Yield

We now investigate practical methods of achieving accuracy in integrated circuits. As discussed earlier, we focus on voltage comparators. We consider comparators in the context of a flash ADC, but accurate comparators are required in most ADC schemes and in several other types of analog circuits.

3.1. Comparator accuracy requirement

To put our discussion of comparator accuracy in perspective, we now review the link between comparator accuracy and the yield of a flash ADC. We first develop expressions for the yield

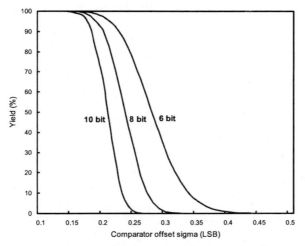

Fig. 3.   ADC yield versus comparator offset in LSB.[16]

of an N bit flash ADC, defining a good ADC as one without non-monotonicity. We use the approach applied by Pelgrom et al.[16]. An N bit flash ADC has 2^N-1 comparators, with the comparator trip-voltages ranging from $V_{trp\_ideal,1}$, the minimum trip-voltage, and increasing in 1 LSB steps to $V_{trp\_ideal,2^N-1}$. In practice, the actual trip-voltages differ from the ideal values. If we assume the error in the trip-voltage is solely due to comparator offset, and if comparator offset has a Gaussian distribution, then we can say:

$$V_{trp,j} = V_{trp\_ideal,j} + \varepsilon, \quad \varepsilon \sim N(0, \sigma_{comp}) \tag{11}$$

In a flash ADC, non-monotoncity occurs when two comparator trip-voltages are interchanged, or in algebraic terms when $V_{trp,j+1} < V_{trp,j}$. (Ideally, $V_{trp,j+1} = V_{trp,j} + LSB$.) If p is the probability that two adjacent comparators become interchanged, or

$$p = P(V_{trp,i+1} < V_{trp,i}), \tag{12}$$

then the yield is the probability that none of the 2^N-2 pairs of the adjacent comparators are interchanged:

$$Yield = (1 - p)^{2^N - 2}. \tag{13}$$

Based on this analysis, in Figure 3 we plot the yield of 6, 8 and 10 bit ADCs versus comparator offset standard deviation expressed in LSBs. We note the standard deviation of comparator offset has to be in the order of LSB/5 to achieve good yield, partly explaining the difficulty in building moderate resolution flash ADCs in CMOS.

In a more strict definition of yield, we define a good ADC as one with a maximum absolute value of differential non-linearity (DNL_{max}), less than a certain value. Ideally, the difference between the trip voltages of adjacent comparators is 1 LSB. For a particular code j the DNL can be defined as:

$$DNL_j = V_{trp,j+1} - V_{trp,j} - LSB$$

(14)

If we specify a maximum allowable DNL, (DNL_{max}) then the probability p of exceeding DNL_{max} at a particular code can be written as:

$$p = P(|V_{trp,j+1} - V_{trp,j} - LSB| > DNL_{max})$$

(15)

The probability that the DNL is not excessive is $1-p$. Since DNL is defined between 2^N-2 pairs of trip voltages, the probability that the ADC is good is:

$$Yield = (1 - p)^{2^N - 2}$$

(16)

The minimum possible value of DNL is -1 LSB, and occurs when there is a missing code. Therefore, the revised definition of yield is only valid for DNL_{max} less than or equal to 1.

3.2.   *Setting transistor area to ensure accuracy*

We begin with one of the simplest approaches, that is, setting transistor size to achieve a certain accuracy and yield. We explore the shortcomings of this approach and go on to examine the use of offset cancelled preamplifiers as an alternative. The use of preamps to achieve accuracy has been very effective. However, we show that preamp power consumption will increase dramatically as CMOS processes evolve. Finally, we propose digital dominant methods of enhancing accuracy. These methods are well suited to nanometer CMOS technologies and break the link between accuracy and power constraints that limit the performance of most analog circuits. Since mismatch is inversely proportional to gate area, we can simply make the transistors large enough to achieve the accuracy we require. However, although device sizing can be used to give stand-alone latching-comparators sufficient accuracy, this approach is inefficient in terms of power and area.

Figure 4 shows an effective and very popular CMOS comparator.[17] The comparator is comprised of a PMOS input differential pair (M_1 and M_2) and NMOS and PMOS cross-coupled latching transistors (M_4, M_5 and M_6, M_7). The comparator is controlled by two

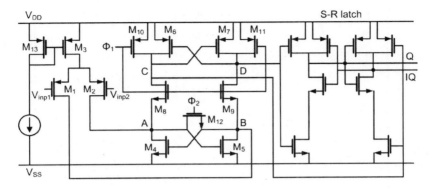

Fig. 4. CMOS comparator.[17]

non-overlapping clocks $\phi 1$ and $\phi 2$. There are three phases of operation. While $\phi 1$ is low and $\phi 2$ is high, switches M_8 and M_9 are open, disconnecting the NMOS and PMOS cross-coupled devices. Pull-up devices (M_{10} and M_{11}) short the output nodes to V_{DD}. At the same time, the reset switch M_{12} is on, shorting the differential voltage developed across the N cross-coupled devices. Since this reset device has a finite on-resistance, the voltage across the switch tracks the input voltage to the comparator. The next phase occurs during the brief non-overlap period. The switch opens and regenerative gain amplifies the input to NMOS cross-coupled devices. During the next phase $\phi 2$ is low and $\phi 1$ is high, and both the NMOS and PMOS cross-coupled pairs become active, producing the comparison output at nodes C and D. The gain during the short non-overlap period reduces sensitivity to offset in the PMOS cross-coupled pair. The SR latch converts the output of the regenerative comparator to CMOS logic levels Q and IQ.

The offset of the devices of the differential pair dominates the input offset, since the mismatch of the N cross-coupled latching devices is divided by the gain during tracking, while the offset of the P cross-coupled devices is further attenuated by the regenerative gain of N cross-coupled latch. The input referred offset of the comparator has a mean of zero, since we assume the transistors have no systematic offset. The random transistor V_{T0} mismatches contribute to the variance of the input referred offset σ^2_{comp}. We can say that[16]:

$$\sigma^2{}_{comp} = k_{comp}{}^2 \sigma^2{}_{\Delta VT0} = k_{comp}{}^2 \frac{A_{VT0}{}^2}{WL}, \tag{17}$$

where W and L are the dimensions of the devices in the input differential pair. The constant k_{comp} depends on the design of the comparator and should have a value greater than 1. We can quickly verify the accuracy reported for this comparator. Given that M_1 and M_2 are 24μm/1.5μm, assuming that the offset of the input pair dominates (i.e. $k_{comp}=1$), and applying

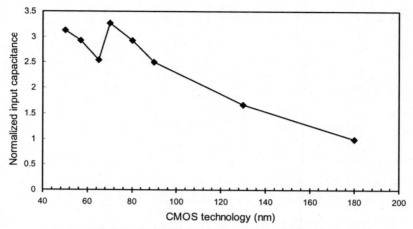

Fig. 5. Normalized input capacitance to achieve constant dynamic range over offset.

the mismatch data presented in Table 1, we estimate σ_{comp} to be 3.5 mV. Considering the 2.5 V input range reported for this 5 V device, we calculate $\sigma_{comp} = 0.18$ LSB at 7 bit resolution. Referring to Figure 3, we predict high yield for a 7 bit flash ADC, but we also note that in practice k_{comp} is somewhat larger than 1.

Although, this method is effective at achieving yield, it is not efficient, particularly for technologies with line widths finer than 1μm. We now consider how the input capacitance varies as we scale to newer CMOS technologies. The input capacitance is important since it determines the amount of power required to drive the comparators. In most cases, the comparators are driven by the output of a sample and hold amplifier or by another on-chip block such as a filter or automatic gain control (AGC) amplifier. We assume a scaling factor of α. Since the power supply voltage scales by α, to achieve the same dynamic range over offset we must also scale σ_{comp} by α. We apply the data in Table 1 to calculate the trend in gate area and gate capacitance to achieve a constant dynamic range over offset. The plot shown in Figure 5 shows that input capacitance grows steadily as line width shrinks.

3.3. *Adding a preamplifier*

By gaining up the differential input to each comparator with the help of a preamplifier, we can relax the input offset requirement of the comparators. A preamplifier gain of A_P increases the permitted offset of the comparator by A_P. This has a very significant effect on the comparator input capacitance, and therefore, also improves power consumption. The addition of a preamplifier provides some other important benefits; in particular, the kickback to the overall input is reduced. Kickback to the input occurs as a comparator latches to a value or is reset.

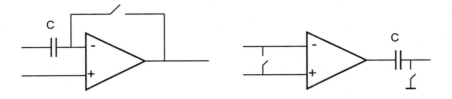

Fig. 6. (a) Input offset storage and (b) output offset storage.

The preamplifier serves to buffer the input from kickback. It is straightforward to calculate the reduction in comparator input capacitance, if we assume that the input devices are the dominant source of offset. Since the offset is inversely related to the square root of gate area, the gate area and also the gate capacitance are reduced by A_P^2.

Not surprisingly, a preamplifier is only effective if the preamplifier itself has low input offset. Several techniques have been developed to control or suppress the offset of the preamplifiers. As with the comparator, the devices in the preamplifier could be made large enough to control offset; however, this approach is not efficient. We concentrate on two of the most popular and effective techniques: spatial filtering and switched capacitor offset cancellation.

One of the most effective techniques is to apply switched capacitor (SC) techniques to store and cancel preamplifier offset. Switched capacitor techniques were made possible by the high input impedance amplifiers and the good switches available in MOS technology. Early on, SC techniques were applied to cancel preamplifier offset and reduce comparator or amplifier offset[18,19,20]. Figure 6 shows two forms of capacitor based offset cancellation: input offset storage and output offset storage.[21] With input offset storage, the amplifier is placed in unity gain feedback and the inputs are shorted to store the offset on the coupling capacitors. After offset cancellation, the overall input offset becomes:

$$V_{os} = \frac{V_{os1}}{1 + A_0} + \frac{\Delta Q}{C}.$$ (18)

Here V_{os1} is the original offset, A_0 is the gain of the amplifier, ΔQ is the mismatch in charge injected by the switches, and C is the capacitor. In contrast, output offset storage works by shorting the amplifier inputs and the storing the output referred offset voltage. With this form of offset cancellation the overall offset becomes:

$$V_{os} = \frac{\Delta Q}{A_0 C}. \tag{19}$$

These equations show that output offset storage completely cancels offset, and in addition, attenuates the charge injection voltage by the amplifier gain. Although, output offset storage seems more effective than input offset storage, it does have some limitations. During offset storage the output voltage becomes the gain A_0 times the input offset. Neither the gain nor the offset can be too large or the amplifier may saturate, failing to store the complete output referred offset on the storage capacitor. For this reason, in practice more than one amplifier stage is often used and the cascade of stages utilizes both input and output offset storage.[20]

Although the SC offset cancelled preamplifier technique is effective, it does have some limitations. Since the gain of each stage is limited, the offset and charge injection at each stage must be carefully considered. Because of leakage, the offset voltage must be restored periodically, preventing continuous operation of the comparator in an ADC. In the long term, as CMOS technology evolves, the advantages of CMOS for SC operation are being eroded. For example, good analog switches are becoming difficult to implement and it is becoming more difficult to achieve amplifier gain.

3.4.   *Spatial filtering of preamplifier outputs*

Spatial filtering[22-24] uses the similarity of the outputs of adjacent preamplifiers to suppress preamplifier offset. This process is illustrated in Figure 7. Since a preamplifier output can be approximated by a combination of the outputs of its neighbors, this information can be used to average out, or filter, random errors. In this example, V_{bL} is the average of V_{aL} and V_{cL}. In practice, averaging is implemented by connecting the outputs of adjacent preamplifiers with resistors (labeled R_1 in the figure).

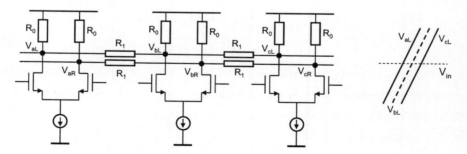

Fig. 7.   Spatial filtering or averaging.

Averaging, or spatial filtering, has a number of advantages over SC offset storage and cancellation techniques, and over the past few years this scheme has largely supplanted SC based techniques in flash ADC design. A significant advantage is that this is a continuous time method. No clock is required, and there is no requirement for periodic reset. However, there are also some drawbacks to this method. As with SC offset cancellation, spatial filtering reduces only preamplifier offset. The overall preamplifier gain must still be large enough to overcome the offset of the latching comparator. Spatial filtering works because the preamplifiers close to the decision level are not saturated. Considering that the ADC LSB voltage is usually far less than $\sqrt{2}\ V_{DSAT}$, spatial filtering operates over a span of several preamplifiers. For this reason, the edge of the comparator string must be terminated with a series of dummy preamplifiers, increasing area and power consumption.

An optimally designed spatial filtering network reduces offsets[24] by a factor of 3. In practice, this means that preamplifier offset is still a consideration. In Figure 8, we plot the variation of input capacitance versus CMOS technology feature size, for 6 and 8 bit flash ADCs. This analysis uses the methodology described earlier to calculate the comparator input offset standard deviation sufficient to achieve a yield of 90%. We assume that offset comes solely from the input devices and the offsets are averaged by an optimal spatial filtering network. The graph shows a significant increase in input capacitance for a 6 bit flash ADC (almost 5x increase from 180 nm to 50 nm); however, the large values of capacitance in the 8 bit case are even more striking.

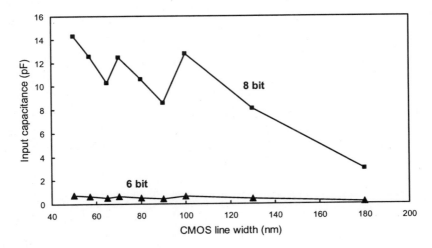

Fig. 8. Preamp input capacitance with optimum spatial filtering.

3.5.   The limitations of preamplifiers

The use of preamplifiers has been an effective method of ensuring accuracy; however, this technique suffers from limitations when applied in nanometer CMOS technologies. Interestingly, we will see a fundamental limitation related to power dissipation. We have seen that switched capacitor offset cancellation, and spatial filtering both have disadvantages. In particular, switched capacitor techniques are not well suited to nanometer CMOS, while spatial filtering is only partially effective at eliminating offset. Nevertheless, our analysis assumes that we can somehow build a perfect preamplifier with no offset. We now show that even with a perfect preamplifier, this technique may become uncompetitive in the nanometer CMOS regime.

We study how preamplifier power dissipation will change as CMOS evolves. To simplify our analysis we assume that there is only one preamplifier gain stage. We model the preamplifier as a common source amplifier with driver transconductance g_m and with resistor loads R_L. (The use of linear resistor loads improves settling behavior.) If the current flowing in the preamplifier is I_A, then g_m is proportional to $2I_A/(V_{GS}\text{-}V_{TH})$, while the power consumption of the preamplifier is $V_{DD}I_A$. For maximum efficiency, the overdrive voltage $V_{GS}\text{-}V_{TH}$ is set at the lowest voltage that robustly ensures saturation, perhaps 200 mV in practice.

We assume that the preamplifier itself has no offset and that the comparator is the only source of offset. We assume that the comparator input offset is dominated by the threshold voltage mismatch of the input differential pair, and as earlier assumed:

$$\sigma^2_{comp} = k_{comp} \frac{A_{VT}^{\,2}}{W_c L_c}. \tag{20}$$

The input capacitance of the comparator is proportional to $C_{ox}W_cL_c$. We consider this capacitance to be the total capacitive load of the preamplifier.

We apply a scaling constant α, so that the power supply voltage and other parameters scale with α. We assume that R_L is chosen to achieve the maximum preamplifier gain for a given transconductance, therefore, the voltage drop across R_L is the maximum allowed that will keep the drive transistors in saturation. Noting that both the MOS saturation voltage and V_{DD} tend to scale with α, we can conclude that the voltage drop across R_L scales with α. The preamplifier gain G_A is given by $R_L g_m$. Since the minimum overdrive voltage (i.e. $V_{GS}\text{-}V_{TH}$) does not scale, we can stay that g_m is proportional to I_A, and, therefore, G_A is proportional to $R_L I_A$. Since the G_A is proportional to the voltage drop across the load, which in turns scales with α, we conclude that preamplifier gain G_A scales with α.

Next we consider offset. To maintain the same dynamic range over random offset the input referred offset of the preamplifier and comparator combination should scale with α, as with the power supply voltage. However, since the preamplifier gain also scales with α, the input offset of the comparator must scale with α^2. Returning to the comparator input offset, as mentioned earlier, there is evidence that A_{VT} scales with t_{ox}, we can say that:

$$\sigma_{comp} \propto \frac{t_{ox}}{\sqrt{W_c L_c}}.$$

(21)

Since t_{ox} scales with α, and we require σ_{comp} to scale with α^2, the gate area must scale with $1/\alpha^2$. The increased gate area has implications on the input capacitance of the comparator, which in turn is the load presented to the preamplifier. The input capacitance is proportional to the gate area:

$$C_c \propto C_{ox} W_c L_c \propto W_c L_c / t_{ox}.$$

(22)

Considering the scaling of t_{ox} and gate area, we see that the comparator input capacitance scales with $1/\alpha^3$. To maintain the same bandwidth, the load resistance of the preamplifier must be reduced to counteract the increase in capacitive load. However, since $I_A R_L$ scales with α, this suggests that the preamplifier current scales with $1/\alpha^2$. Finally, since we now know how the preamplifier current and supply voltage scale, we can conclude the preamplifier power dissipation scales with $1/\alpha$. Figure 9, based on actual and predicted values of mismatch parameters and transistor characteristic, confirms this trend of increasing preamplifier power consumption with process scaling.

We note that this analysis ignores the power consumption of the comparator itself. Our analysis suggests that comparator input offset must scale with α^2, to counteract the reduction in power supply voltage and the associated reduction in preamplifier gain. We assume that comparator input offset is proportional to the mismatch of the comparator input differential pair, and that the mismatch of the input pair dominates. We now look at the trend in power consumption of the comparator itself, concentrating on the cross-coupled latching transistors. Using an analysis similar to that presented Uyttenhove and Steyaert,[9] we arrive at an *optimistic* baseline analysis of the power consumption trend, by assuming that the accuracy of these MOSFETs should remain constant relative to supply voltage or $\sigma_{latch} \, \alpha \, V_{DD}$.

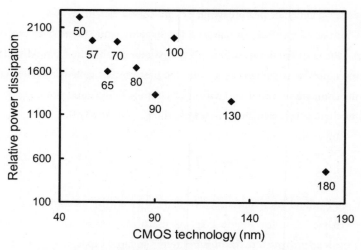

Fig. 9. Relative preamplifier power dissipation for constant bandwidth.

From our analysis of mismatch (Eq. 4) we have:

$$W_L L_L \propto \frac{A_{VT0}^2}{V_{DD}^2}.$$ (23)

Ignoring capacitances other than gate capacitance, the speed of the latching devices is proportional to g_m/C_{GS}. Substituting for g_m and C_{GS}, we can say:

$$\frac{I_L}{speed} \propto W_L L_L C_{ox}\left(V_{GS} - V_{T0}\right),$$ (24)

where I_L is the comparator current. Since comparator power is proportional to $I_L V_{DD}$, and substituting for $W_L L_L$ (Eq. (22)), we find that:

$$\frac{Power}{speed} \propto A_{VT0}^2 C_{ox} \frac{V_{GS} - V_{T0}}{V_{DD}}.$$ (25)

When we consider the first order scaling trends for C_{ox}, A_{VT}, and V_{DD}, we see that power dissipation remains constant for a given speed. However, when other factors, such as drain bulk capacitance are considered, the power consumption increases with scaling.[9]

Although, this analysis makes many simplifications, it is clear that with the current design styles, scaling will ultimately lead to increased power consumption. Our analysis assumes a single amplifier stage for simplicity. Adding extra preamplifier stages increases the overall preamplifier gain, but this approach also leads to increased power consumption, especially when the offsets of the preamplifier stages are considered.

Our analysis does not consider velocity saturation. Since velocity saturation tends to reduce transconductance, our analysis understates the increase in power consumption. Furthermore, the assumptions about the trends in MOS matching are also optimistic. It is not likely that A_{VT0} will continue to decrease with scaling, and in addition, the current factor mismatch may also become more important. Both of these factors will also argue for a more pronounced increase in power consumption.

4. Alternative Techniques for Nanometer CMOS

For digital circuits, each process generation delivers higher gate density, more speed, better power efficiency, and lower cost per gate, than its predecessor. But this evolution is accompanied by a deterioration of 'analog' characteristics. We now consider two techniques, *trimming* and *redundancy*, that take advantage of the digital performance of nanometer CMOS. Because of their complexity, these techniques are not attractive in older CMOS technologies but become advantageous in nanometer CMOS because of the reduced area and power required for digital processing. Redundancy and trimming greatly relax the analog accuracy requirements of comparators and other analog circuits, and help decouple the traditional constraints of accuracy and speed.

4.1 *Trimming*

In trimming schemes, a trim current (or voltage) is applied to cancel the offset of each comparator (or preamplifier). Pelgrom et al.[16] describe one analog trim technique. More recently, digitally-controlled trim techniques have been presented.[25] The trim value for each comparator is stored in a register and converted to current (or voltage) with separate DACs. A calibration routine, initiated at power-up, programs the appropriate value in each register. The DAC LSB size is directly related to the required DNL.[26] For example, the DAC LSB size should be less than that corresponding to a 1 LSB change in offset in order to achieve $|DNL| <$ *1* LSB. The DAC resolution and the size of the register that controls it, depends on the statistics of comparator offset and on the accuracy of the DAC. The range of the DAC depends on the variability of comparator offset and the required probability that the calibrated comparator offset should achieve a particular DNL (i.e. the yield). This probability p that the

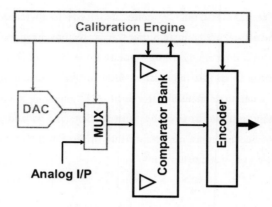

Fig. 10.   Flash ADC with comparator redundancy.[27]

DAC range must accommodate the worst comparator offset is related to the required ADC yield:

$$Yield = p^{2^N - 2} \tag{26}$$

For large values of σ_{comp} this scheme can have a significant analog and digital hardware overhead. For example, if we require a 6 bit ADC to have 99% yield, then p must be at least 0.9998. If we assume a Gaussian distribution for comparator offset, this implies a DAC range of +/- 3.7 σ_{comp}. If $\sigma_{comp} = 3$ LSB, and we want to calibrate to $|DNL| < 0.5$ LSB, then each trim DAC should have a resolution of almost 6 bits.

4.2. *Redundancy*

Instead of building or designing components to achieve high accuracy, an effective alternative is to build several lower accuracy components and select the most suitable component from a set of redundant components.[26,27] We rely on *redundancy*, and not on tight tolerance to achieve accuracy. Since the constraint of accuracy is removed, circuit speed and accuracy are decoupled.

In a flash ADC, redundancy can be used to overcome comparator offset. Instead of one comparator per code (or 2^N-1 in total), R comparators are assigned to each code (i.e. $R(2^N-1)$ comparators in total). At startup, the 2^N-1 most suitable comparators are selected. For each code, a selection routine searches for the comparator whose trip-voltage is closest to the ideal trip-voltage, with the aid of a DAC. It should be noted that the comparator chosen to represent a code may not have been nominally assigned to that code. For example, a comparator nominally assigned to code 8, may be chosen as the comparator to represent code 10, if it has an offset of exactly +2 LSB. Figure 10 is a block diagram of a flash ADC with redundant comparators.[27] In this redundancy scheme, the unused comparators are powered down, and only 2^N-1 comparators dissipate power.

Simulations indicate that excellent yield is achieved with four comparators per code. It should be noted that the overall comparator area is only a fraction of the area required to achieve the same result by simply relying on transistor size to achieve low offset. This is illustrated in the results from a Monte-Carlo simulation shown in Figure 11. In the figure, the simulated effective resolution of a 6 bit flash ADC with redundancy is compared with that of a conventional ADC. In this comparison, the conventional ADC marked *4x size comps* has the same *total* comparator area as the ADC with redundancy. With σ_{comp} = 3 LSB, the ADC with redundancy has a median SNDR of 5.7 effective bits, while the corresponding SNDR for a conventional ADC of the same total comparator area is only 4.2 effective bits.

An example helps us to appreciate the power savings that can be achieved with the help of redundancy. In this example, we compare two 6 bit ADCs, one with and one without redundancy. If we use a redundancy of 5 comparators per code, even with a very large comparator offset of σ_{comp}=5 LSB, we achieve a yield of 98.9%.[26] We now compare the comparator area to that of a conventional flash ADC, where transistor size alone is used to assure accuracy. Assuming that the offset error comes solely from transistor mismatch, and that this mismatch is inversely related to gate area (as is the case with V_{TH} mismatch), the comparators must be 600 times larger to achieve the same yield. (Overall, the conventional ADC is 120 times larger than the one with redundancy.) If we assume that speed is related to

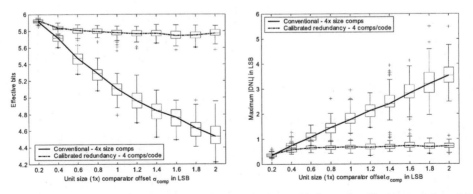

Fig. 11. Comparison of 6 bit ADC with and without redundancy; (a) shows the effective resolution while (b) compares the worst case DNL.[26]

g_m/C and that transconductance is proportional to current, then the example shows a factor of 600 difference in power consumption for a given speed.

Although redundancy is very effective in flash ADCs, it can be applied to comparators in any application. In the same way, redundancy can be applied to any circuit where there is a tradeoff between accuracy (related to component mismatch) and speed. We note that it is the selection of a component from a bank of redundant devices that dramatically improves resilience to mismatch. Selection introduces new information that improves accuracy. With this technique, accuracy is far better than can be achieved by simply increasing component area.

Acknowledgements

This work was supported in part by the National Science Foundation (Award CCF-0346874) and by Intel. The authors also thank Dr. Dennis Sylvester, Dr. Linda Sattler and Conor Donovan.

References

1. P. Kinget and M. Steyaert, "Impact of transistor mismatch on the speed-accuracy-power trade-off of analog CMOS circuits", Custom Integrated Circuits Conference (1996) 333—336.
2. M. Chen, J. Ho and T. Huang, "Dependence of current match on back-gate bias in weakly inverted MOS transistors and its modeling", IEEE J. SSC, vol. 31, no. 2 (Feb 1996) 259—262
3. F. Forti and M. Wright, "Measurement of MOS current mismatci in weak inversion" IEEE J. SSC, vol. 29, no. 2, (Feb 1994) 138—142
4. J. Bastos, M. Steyaert, A. Pergoot, W. Sansen, "Influence of die attachment on MOS transistor matching" IEEE Transactions on Semiconductor Manufacturing, vol. 10, issue 2 , (May 1997) 209-218.
5. H. Tuinhout and M. Vertregt, "Characterization of systematic MOSFET current factor mismatch caused by metal CMP dummy structures" IEEE Transactions on Semiconductor Manufacturing, vol 14, issue: 4, (Nov. 2001) 302-310.
6. S. Lee and D. J. Allstot, "Techniques for fast electro-thermal simulation of ICs," IEEE International Solid-State Circuits Conference, vol. 36, (Feb 1993) 120 - 121.
7. M. J. M. Pelgrom, C. J. Duinmaijer, and A. P. G. Welbers, "Matching Properties of MOS Transistors", IEEE J. Solid State Circuits (Oct 1989) 1433—1440.
8. J. Shyu, F. Krummenacher, and G. C. Temes, "Random error effects in matched MOS capacitors and current sources," IEEE Journal of Solid-State Circuits, vol. 19, (Dec 1984) 948 – 956.
9. K. Uyttenhove M. S. J. Steyaert, "Speed-Power-Accuracy Tradeoff in High-Speed CMOS ADCs", IEEE Transaction of Circuits and Systems-II (Apr 2002) 280—287
10. Process Integration, Devices, and Structure, International Technology Roadmap for Semiconductors (2003) (www.itrs.org)

11. P. Kinget and M. Steyaert, "Impact of transistor mismatch on the speed-accuracy-power trade-off of analog CMOS circuits", Custom Integrated Circuits Conference (1996) 333—336.
12. A. Hori, H. Nakaoka, H. Umimoto, K. Yamashita, M. Takase, N. Shimizu, B. Mizuno, and S. Odanaka, "A 0.05μm-CMOS with ultra shallow source/drain junctions fabricated by 5keV ion implantation and rapid thermal annealing", International Electron Device Meeting (1994) 485—488.
13. P. A. Stolk and D. B. M. Klaassen, "The effect of statistical dopant fluctuations on MOS device performance", International Electron Devices Meeting (1997) 627—630.
14. M. J. M. Pelgrom, H. P. Tuinhout, and M. Vertregt, "Transistor matching in analog CMOS applications", International Electron Device Meeting (1998) 915—918.
15. H. Tuinhout, "Impact of parametric mismatch and fluctuations on performance and yield of deep-submicron CMOS technologies", 32th European Solid-State Device Research Conference (2002).
16. M.J. Pelgrom, A.C.J. v. Rens, M. Vertregt, M.B. Dijkstra, "A 25-Ms/s 8-bit CMOS A/D converter", *IEEE J. SSC*, vol. 29, pp. 879-886, Aug. 1994.
17. G. M. Yin, F. O. Eynde, and W. Sansen, "A High-Speed CMOS Comparator with 8-b Resolution", IEEE J. Solid-State Circuits (Feb 1992) 208—211.
18. R. Poujois, B. Baylac, D. Barbier, and J. Ittel, "Low-Level MOS transistor amplifier using storage techniques", IEEE International Solid-State Circuits Conference (1973) 152—153.
19. D. A. Hodges, P. R. Gray, R. W. Brodersen, "Potential of MOS technologies for analog integrated circuits", IEEE J. Solid-State Circuits (Jun 1978) 285—294.
20. D. J. Allstot, "A Precision Variable-Supply CMOS Comparator", IEEE J. Solid-State Circuits (Dec 1982) 1080—1087.
21. B. Razavi and B. A. Wooley, "Design Techniques for High-Speed, High-Resolution Comparators", IEEE J. Solid-State Circuits (Dec 1992) 1916—1926.
22. H. Pan, M. Segami, M. Choi, J. Cao, and A. A. Abidi, "A 3.3-V 12-b 50-MS/s A/D converter in 0.6-μm CMOS with over 80-dB SFDR," IEEE Journal of Solid-State Circuits, vol. 35, pp. (Dec 2000) 1769 - 1780.
23. K. Kattmann and J. Barrow, "A technique for reducing differential non-linearity errors in flash A/D converters," IEEE International Solid-State Circuits Conference, vol. 34, pp. (Feb 1991) 170 - 171.
24. H. Pan and A. A. Abidi "Spatial filtering in flash A/D converters" IEEE Transactions on Circuits and Systems II, vol 50, no. 8, (Aug 2003).
25. Y. Tamba and K. Yamakido, "A CMOS 6-b 500Msample/s ADC for hard disk drive read channel", IEEE International Solid-State Circuits Conference (Feb 1999) 324-325.
26. M. P. Flynn, C. Donovan, and L. Sattler, "Digital Calibration Incorporating Redundancy of Flash ADCs", IEEE Transactions on Circuits and Systems-II (May 2003) 205-214.
27. C. Donovan and M. P. Flynn, "A "digital" 6-bit ADC in 0.25-um CMOS", IEEE J. Solid-State Circuits (Mar 2002) 432-437.

International Journal of High Speed Electronics and Systems
Vol. 15, No. 2 (2005) 277–295
© World Scientific Publishing Company

SELF-INDUCED NOISE IN INTEGRATED CIRCUITS

RANJIT GHARPUREY

Department of Electrical Engineering and Computer Science, University of Michigan
1301 Beal Avenue, Ann Arbor, Michigan 48109, United States of America
ranjitg@eecs.umich.edu

SHAHRZAD NARAGHI

1301 Beal Avenue, Ann Arbor, Michigan, United States of America
naraghi@eecs.umich.edu

Circuits with diverse electrical behavior are often placed in close physical proximity in order to achieve high-levels of on-chip integration. The activity of certain types of circuits can generate harmful interference, and degrade the performance of the system through electromagnetic coupling. Considerable effort in system-on-a-chip implementations is in fact related to technology and architectural considerations for minimizing this interference. This is especially the case in systems that have exacting requirements on the dynamic range such as those for wireless applications.

In this paper, we will discuss the evolution of techniques for modeling and analyzing these sources of noise generation and interference. We will provide a physical description of the problem. Techniques for extraction of electrical models to represent the media that support these noise sources will be covered. Macromodeling techniques will be discussed. Finally we will introduce the concept of functional modeling of circuit functions and present such a model for an integrated flash analog-to-digital converter.

Keywords: Self-induced, substrate, package, coupling, noise, modeling, functional-modeling

1. Introduction

The recent years have seen a proliferation of systems for various applications that require a mixed-signal implementation, where sensitive analog or small-signal circuits are placed in close proximity to large digital cores. Examples include systems for communications and consumer electronics. This is a departure from earlier systems for similar applications that were primarily analog in nature. The increase in digital content of systems is a natural outcome of the progressively shrinking dimensions of commercial CMOS technologies. Designing systems to be digital-intensive allows for several well-known benefits in terms of cost, performance and robustness of implementations. With increasing digital content, these systems can benefit from scaling advantages, arising from Moore's Law[1], that have allowed for improvement in performance and integration levels in computing systems over several decades. For example, communication systems for mass applications, such as cellular telephony have seen a continual improvement in the bit-rate per unit cost and bit-rate per unit power over the past decades. To a significant extent this is due to the increasing use of digital signal processing in the implementations. Unlike computing applications however, many of these systems will continue to require a significant analog front-end in order to interface to the sources of information such as speech and video, which tend to be inherently continuous time and continuous amplitude. Thus, these systems will retain their mixed-signal character even though the digital content will increase.

The input and output signals of analog and digital circuits have very different characteristics in frequency and time. Digital signals typically have a frequency content that spans a large band, and are characterized by abrupt transitions in the time-domain. Most analog functions on the other hand tend to have relatively limited spurious harmonic content and smooth time-domain characteristics. These circuits also require very different operating environments. While digital circuits can operate fairly robustly in the face of spurious environmental noise, analog circuits that process signals with a continuous time and amplitude information can experience an irreversible degradation in the quality of the output signal in the presence of externally introduced noise. In mixed-signal applications, the activity of digital circuits often generates signals that appear like noise to analog circuits and thus tends to degrade the performance of these circuits.

We refer to the above noise signals that are related to circuit activity as self-induced noise in the following sections of this paper. Current flow through the substrate was first recognized as a potential hazard by engineers concerned with the latch-up and other reliability problems [2,3]. Later, with the growing number of mixed-signal designs, it became apparent that noise injected by digital circuits could limit the performance of analog circuits and could potentially be one of the major sources of degradation in IC performance [4]. Parasitics associated with the physical IC substrate, interconnects, and the IC package were recognized as significant contributors in the coupling of digital switching activity into the analog portions of the IC.

In this paper, we will provide an overview of several techniques for modeling of self-induced noise. We will start with a description of the basic mechanisms by which noise is generated and coupled into the substrate or interconnect impedance. Reception mechanisms, by which the noise is sensed in the analog portions of the IC are also discussed. There has been considerable research in the recent years on modeling this phenomena, starting from techniques for efficient extraction of substrate and interconnect parasitics, to macromodeling the signatures of digital gates for simulation of parasitic noise at a higher level of abstraction. Some of these techniques will be discussed. Finally we will introduce the concept of functional level modeling of self-induced noise.

The recent years have seen remarkable advances in integration levels in many high-dynamic range systems for wireless communications, such as cellular telephony, wireless

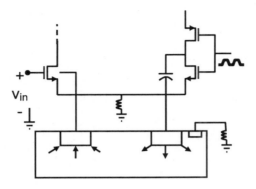

Fig. 1. Substrate noise and ground-bounce in a mixed-signal IC

networking, and sensor applications. Similar advances have also been seen in broadband wired systems and high-speed serial transceivers. A careful understanding and management of self-induced noise sources, has enabled much of this integration. However, emerging technologies in several of these application spaces, promise new challenges in this area. We expect, for example, that in broadband wireless systems, such as Ultra Wideband, special consideration will have to be paid to self-induced noise sources. It is logical to expect that this interference will become a greater problem in such systems, since a greater portion of the spectrum of self-induced noise will overlap with the input signal, and spectral-domain isolation will thus be challenging. It should be noted that spectral domain isolation techniques have been a key tool for system architects to enhance integration in current day narrow-band systems where the signal bandwidth is often a very small fraction of the carrier frequency. It is even conceivable that the sensitivity in broadband wireless transceivers will be set by self-induced noise rather than thermal noise limits. Thus, this problem will potentially be a greater issue in new generations of wireless technologies. As will be discussed later, our work on functional modeling of self-induced noise is motivated by the need for accurate estimation of sensitivity limits in broadband wireless transceivers.

2. Self-Induced Noise–Injection and Sensing Mechanisms

The primary generators of self-induced noise are circuits that process large input and output signals, often of a switching nature. We refer to these circuits as noise generators. Digital CMOS gates are a typical example. Non-digital examples include integrated oscillators and mixers. Circuits that are sensitive to the signal coupled by the noise generators typically process small signals, for example, low-noise amplifiers, or signals that are required to have a high purity, such as oscillators. The devices in the generators couple noise into interconnects and the substrate through various mechanisms[5,6,7,8]. Victim circuits sense noise through other device mechanisms and are summarized in this section.

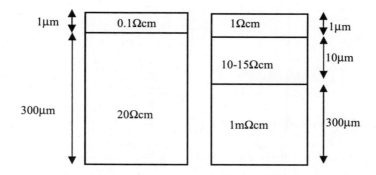

Fig. 2. Typical high-resistivity and low-resistivity substrate cross-sections

Noise coupling can occur due to current flow through non-ideal ground-path impedance [5,6,10,11], which is typically referred to as ground-bounce. If this ground connection is shared by other circuits, current flow in the noise generator will cause spurious coupling into the victim. A rather extreme case of this is shown in Fig. 1., where an inverter is connected to a package pin with a large ground-path inductance. A common source amplifier shares the local internal ground. The input to the common source amplifier is applied externally and is referred to an external ground. Clearly, any local switching activity will be coupled into and amplified by the common source amplifier.

Another injection mechanism that is illustrated in the same figure is a capacitive coupling path from the digital output node to the substrate. Substrate coupled noise can be detected by coupling capacitors (junction and interconnect) at the output and the input of the common source amplifier. This mode of coupling is heavily influenced by the composition and the impedance offered by the substrate. In the ideal case, for example, if the substrate impedance is infinitely large, there will not be any coupling through the substrate itself. The body node of the common-source amplifier will also sense this noise and will couple the noise to the output due to threshold voltage modulation (body effect). This mode of coupling will be especially effective if the body connection is far from the device. Typical types of silicon substrates (low-impedance and high-impedance bulks) are shown in Fig. 2.

In addition to capacitive coupling, noise injection into the substrate impedance can also occur due to hot-electron injection effects, which are prominent in short-channel devices [8,9]. The latter mechanism is inherently rectifying in nature[6], since carrier injection always occurs into the substrate and thus the polarity of the current flow into the substrate is always the same, regardless of the polarity of the transition at the input of the digital gate.

In a switching CMOS inverter, hot-electron induced current will be injected into the substrate during both 0-1 and 1-0 transitions, while the capacitive component of the current will reverse direction during the two edges. As a consequence, hot-electron induced currents will possess primarily even- harmonics of the fundamental switching frequency and a DC component, while the capacitive currents will possess mainly odd-harmonics, and no DC component. The presence of a DC component in any substrate

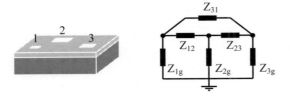

Fig. 3. A three contact substrate problem and equivalent lumped representation

current can be potentially very harmful to circuit operation. In addition to causing a drift in threshold voltages, it can also lead to an increase in minority-carrier injection into the substrate due to partial forward-biasing of device-to-substrate junctions. Further, while fundamental and odd harmonics of capacitively coupled signals can be cancelled to the first order at the substrate node or at the ground or supply impedance by the use of differential circuits, hot-electron currents are not similarly cancelled in differential circuits. The dependence of this current on device bias is given by

$$I_{sub} = K_1 (V_{ds} - V_{ds,sat}) I_D \exp[-K_2 / (V_{ds} - V_{ds,sat})] . \tag{1}$$

where I_D is the drain current, V_{ds} is the drain-to-source voltage and $V_{ds,sat}$ is the drain-to-source voltage at saturation. K_1 and K_2 are semi-empirical constants.[8]

Bipolar transistors sense and inject noise primarily through collector-substrate junction capacitors. Passive devices, such as capacitors and inductors are also coupled to the substrate through parasitic capacitors, through which noise can be injected or sensed based on the circuit configuration.

Most of the above injection and sensing mechanisms are modeled in circuit simulators such as SPICE. The substrate itself on the other hand is not modeled, and must be explicitly modeled. Electrically, for majority carrier flow, the substrate behaves as a distributed impedance between devices. Currents injected into the bulk traverse varied paths, depending on the material constants such as bulk resisitivity and dimensions and also on specific boundary conditions set by the devices and contacts on the surface. In the following section we will discuss techniques to analyze and model this current flow.

3. Extraction of Substrate Parasitics

Two classes of techniques have been used for extracting substrate models. One set of techniques use numerical approaches[12,13,14,15] such as the finite-differences method. The other class uses a semi-analytical approach, for example, those based on integral equation technique[12]. In the low-frequency limit, where inductive effects can be neglected, the problem involves a solution for the LaPlace equation in the substrate, that is:

$$\nabla^2 \phi = 0 . \tag{2}$$

where ϕ is the potential in the substrate. Devices are treated as equipotential contacts in both methods and a lumped resistive or a frequency-dependent impedance network is extracted between these contacts (see Fig. 3). The impedance network can be used in a circuit simulator to model the substrate.

3.1. *The method of finite differences*

In the finite differences approach, the substrate is reduced to a mesh of resistors. By applying a unit potential on any one surface node, connecting all other surface contacts to ground and determining the current through the individual ground contacts, the contact-to-contact impedance can be extracted. This requires the solution of a sparse equation of the form

$$A\phi = b .\tag{3}$$

where the elements of $[A]$ represent the node-to-node conductance in the mesh and $[b]$ is the source voltage vector. Several techniques exist for the efficient solution of the LaPlace equation numerically. In typical problems however, the number of nodes in the mesh can be a prohibitively large number, and the numerical extractor can be slow.

3.2. *Integral equation techniques*

Integral equation based techniques solve for the potential in the medium for an arbitrary charge distribution, by employing a spatial convolution of the charge distribution with the Green function of the medium [16,6,17], which is the potential generated by a point charge.

The Green function is determined by the material constants of the layers that compose the substrate and geometry of the substrate, including the dimensions and the thickness of various substrate layers (Fig. 4). Green functions for substrates have been evaluated in rectangular[16,6,17] and cylindrical coordinates[18]. Given the rectangular dimensions of silicon die, the former solutions are more representative, and can include edge effects more accurately. On the other hand, the solution in cylindrical coordinates is more compact, since it requires two coordinates (radial distance r, and vertical dimension z), unlike the solution in rectangular coordinates. Further, the Green function in the rectangular coordinate system is a two dimensional series, where the coefficients of the series decay gradually with the series index, and requires computation to very high summation indices as shown below.

$$G(\vec{r},\vec{r}') = \sum_m \sum_n f_{mn}(z,z')Cos(m\pi x/a)Cos(n\pi y/b)Cos(m\pi x'/a)Cos(n\pi y'/b) .\tag{4}$$

(a,b) are the lateral dimensions of the substrate, r is the coordinate of the observation point, and r' is the coordinate of the source point charge. The functions f_{mn} at the surface are given by

$$f_{mn}(0,0) = [C_{mn}/ab\gamma_{mn}\varepsilon_N]\times[(\beta_N \tanh(\gamma_{mn}d)+\Gamma_N)/(\beta_N+\Gamma_N \tanh(\gamma_{mn}d))] .\tag{5}$$

where β_N and Γ_N are coefficients arising from the application of potential and normal displacement continuity at the interfaces of the substrate layers and are computed in the surface (N-th) layer. The coefficients γ_{mn} are given by $((m\pi/a)^2+(n\pi/b)^2)^{(1/2)}$. β_N and Γ_N can be computed recursively starting from the coefficients in the bottom layer, given by $\beta_0 (1)$ and $\Gamma_0 (0)$, using

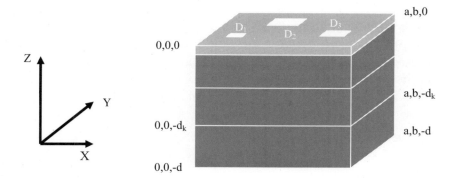

Fig. 4. Multilayered approximation of an IC substrate

$$\begin{bmatrix} \beta_k \\ \Gamma_k \end{bmatrix} = \left[A(\gamma_{mn}, d_k, d) \right] \begin{bmatrix} \beta_{k-1} \\ \Gamma_{k-1} \end{bmatrix}. \tag{6}$$

[A] is a 2x2 matrix with elements that are functions of γ_{mn}, the interface coordinate d_k and the substrate depth d. Using an FFT to compute the Green function provides a very efficient scheme for pre-computation [17]. By employing an FFT, the substrate surface is effectively discretized to a grid of rectangular panels, where the unit panel dimension in a given direction is given by the dimension of the substrate along that axis, divided by the number of FFT computation points. By use of the FFT, the potential on any panel j, due to a unit charge evenly distributed on any other panel i (p_{ij}), is pre-computed. This coefficient is related to the Green function by

$$p_{ij} = \frac{\phi_i}{Q_j} = \frac{1}{S_i S_j} \int \int G(r, r') ds' ds. \tag{7}$$

where S_i and S_j represent the surface area of unit panels i and j respectively.

It should be pointed out that the coefficients β and Γ should not be evaluated using the above notation. If the above recursion is used, β_N and Γ_N rapidly converge to 1 and -1 respectively for large values of m and n, leading to a numerically unstable form (0/0) for p_{ij}. A numerically stable technique for computation of these terms was used in [6], the sub-routines for which are shown below.

```
/*kr[m1] = ε[m1-1]/ε[m1]*/
/*m1 is the Layer number*/
/*dz[m1] is the z coordinate of the bottom of layer m1*/
double bgnr(m1,m2,gama)
  {double x,bgdr();
    if(m1==0) return(tanh(gama*(dz[0]-dz[1])));
    else
```

```
        {
    x=tanh(gama*(dz[m1]-dz[m2]));
    x*= (kr[m1]*bgdr(m1-1,m1,gama));
    x += bgnr(m1-1,m1,gama);
    return(x);
        }
}
double bgdr(m1,m2,gama)
  {double x,bgnr();
  if(m1==0) return(1.0); else
        {
        x=tanh(gama*(dz[m1]-dz[m2]));
        x*=bgnr(m1-1,m1,gama);
        x+= (kr[m1]*bgdr(m1-1,m1,gama));
        return(x);
        }
}
```

The routines *bgnr* and *bgdr* represent the numerator and the denominator of Eq. (5) in a recursive layer-wise calculation. The above routines can be shown to be inherently stable. The indeterminate form that arises in a direct computation of Eq. (5) is avoided here.

Physical intuition behind the stable recursion is presented below. The recursive technique builds the solution one layer at a time. We start with the bottom layer (Fig. 4) where the potential is of the form

$$\phi = Sinh(\gamma_{mn}(d+z)) . \tag{8}$$

In the evaluation of the potential for $(-d_1 < z < -d_2)$, we treat the interface at $z = -d_1$ as an equipotential layer, with a potential of $\phi = Sinh(\gamma_{mn}(d-d_1))$, and maintain the continuity of the electric displacement vector, which at $z = -(d_1+\delta)$ is given by $D = \varepsilon_0 \gamma_{mn} Cosh(\gamma_{mn}(d-d_1))$. The general expression for the potential in this layer is given by

$$\phi = \alpha Sinh(\gamma_{mn}(d_1+z)) + \beta Cosh(\gamma_{mn}(d_1+z)) . \tag{9}$$

Equating the potential and the electric displacement at $z = -d_1$, we get,

$$\alpha = (\varepsilon_0 / \varepsilon_1)(Cosh(\gamma_{mn}(d-d_1))); \beta = Sinh(\gamma_{mn}(d-d_1)) . \tag{10}$$

We can then evaluate the potential for $(-d_2 < z < -d_3)$ in a similar manner, by treating the boundary at $z = -d_2$ as an equipotential surface and equating the potential and the normal displacement at this interface. In fact, it can be easily shown that the general expression for the potential in the layer $(-d_k < z < -d_k+1)$ is given by

$$\phi = (D(-d_k) / \varepsilon_k)Sinh(\gamma_{mn}(d_k+z)) + \phi(-d_k)Cosh(\gamma_{mn}(d_k+z)) . \tag{11}$$

where $\phi(-d_k)$ is the potential and $D(-d_k)$ is the magnitude of the electric displacement at $z = -d_k$. For evaluating the potential in any given layer, the layer below it is equivalently replaced by its boundary potential. If the potential evaluated at the lowest layer is finite, the recursive nature of this procedure ensures that the potential in all layers can be evaluated precisely, without numerical instability.

Symmetry of the Green function in the observation layer at the surface between the observation point and the charge location (z,z') allows us to express the above expressions in a recursive form that is equivalent to that in Eq. (5). Numerically stable forms for contacts defined in sub-surface layers of the substrate can also be evaluated in [19].

The average potential over larger panels due to a charge placed on other panels can similarly be efficiently determined by using FFT based computation. Larger contacts are mapped onto these panels. In order to compute the contact-to-contact impedance using this technique, a coefficient of potential matrix is set up. The elements of this matrix represent the potential induced on the panels composing the contacts by a unit charge on a panel within another contact. In order to reduce the size of this matrix, an efficient strategy is to employ unit panel sizes at the periphery of the contacts, where the field is the highest and use progressively larger panel sizes towards the interior of the contacts. The contact-to-contact impedance can be obtained by inverting the coefficient of potential matrix.

The size of the matrices involved in this technique is typically much smaller than those generated using numerical finite-difference based approaches. This is because only the surface contact areas are included in these matrices, instead of the entire substrate bulk. For this reason, this approach is also referred to as the boundary element method [16].

There are two well-known problems that plague this approach. The first is that the matrices generated in this method are dense. Using direct techniques such as LU factorization becomes rapidly impractical with the size of the matrix N, since these techniques are $O(N^3)$. Heuristics to sparsify this matrix have been attempted using various types of relaxation techniques, by the use of wavelet operators, or Krylov subspace iterative methods, such as GMRES, in conjunction with either fast multipole algorithms or eigen-decomposition based techniques[20]. However, accuracy loss in the former and convergence performance in the latter are still major concerns. An efficient formulation along the lines of a hierarchical SVD algorithm for sparsifying the potential matrix, without significant loss in accuracy has also been proposed[21].

Multilevel iterative solvers have also been employed, whereby the algorithm operates on two or more discretizations, from coarse to fine for a given contact configuration. In a two-level formulation[22], at the lower level a number of sub-matrices are locally solved and the resulting residual is projected to the upper level. At the upper level the coarse grid problem is solved explicitly and the result is projected back to the finer problem, where it is used as a starting point for the next iteration. The iteration ends when the residual norm falls below a tolerance.

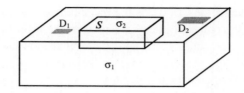

Fig. 5. Lateral resistivity variation in the surface layer

The second problem with the Green function based approach in rectangular coordinates arises due to the requirement for uniform layer resistivity in the lateral directions. Local variations in resistivity are, therefore, difficult to include. A possible solution is to combine a fully numerical approach with the integral equation technique[6]. Consider the surface layer shown in Fig. 5, where we would like to determine the impedance between two contacts D_1 and D_2. The region has a uniform conductance σ_1, except in the volume S, where the conductance is σ_2. A possible method to approach this problem is to compute the Green function in the surface layer assuming a uniform layer resistivity, and computing explicitly the resistors from contacts 1 and 2 to panels on the sides and bottom of the volume S. The volume S itself is modeled by a discrete mesh using a finite difference technique, although with an effective conductance of σ_2-σ_1 and the inter-panel resistors are computed for the volume S. In the limit when the number of panels on the exterior of S tends to infinity, the resistance between contacts D_1 and D_2 can be determined accurately by placing the inter-panel resistors computed using finite-differences on S in parallel with those computed using the integral equation approach for the uniform substrate. Since in practical cases the number of panels will be finite, the accuracy of this technique will be limited in practice. However, this technique does allow for the inclusion of lateral resistivity variation, while minimizing the required finite-difference solutions. Green functions that are required for the computation of the sub-surface contact layers in this approach are described in [19].

Schrik et al. have proposed an algorithm that uses an Integral Equation approach for the bulk and a numerical approach for the surface layer[23]. This approach is much faster and memory-efficient than using a fully numerical approach over the entire bulk. However, the complexity in the surface layer is similar to a finite differences approach.

Full substrate impedance extraction using a numerical, analytical or a hybrid approach has two important issues that need to be considered. The first is the computation time required for substrate extraction mentioned above. The second issue relates to efficient use of the extracted models. An N contact problem leads to $O(N^2)$ simulation elements in the circuit simulator. For a practical IC, it is not feasible to attempt a full scale simulation of this order. Approximations can be employed to partition the problem into smaller sections. Coarse-fine approximation techniques can be employed where the impact of distant contacts is not considered in detail [6,7]. Ultimately, the efficient use of such an extraction tool will rely on user intuition and experience.

3.3. *Noise macromodeling*

Since the use of extraction techniques is restricted to small-to-moderately sized problems, a different approach is required to estimate self-induced noise in larger circuits. Macromodeling of substrate contact impedances has been proposed for this purpose. These techniques can provide a good estimate of substrate parasitics for problems of a limited size[24,25]. A fundamentally different and elegant approach to estimating self-induced noise based on macromodeling of the digital signatures of individual gates has also been proposed[26,27]. The noise injected into the substrate by each gate, for all combinations of input transitions, is simulated by using an extraction based simulator. These waveforms are called noise signatures. Based on the inputs to the gates, an event-driven simulator is used to compute the occurrence of each noise signature for all the gates in the circuit. A noise signal is deduced as a superposition of these noise signatures, for a given set of inputs. The spectrum of the noise injected into the substrate is determined from a discrete Fourier Transform of this signal.

The macromodeling approach has been similarly employed more recently[28,29,30]. These approaches although elegant, are analytical by nature rather than predictive. The substrate noise generated by various input vectors can be deduced only after the digital portion of the IC has been designed. A capability that is very desirable in the design and definition of systems is the ability to predict *a priori* the nature of the self-induced noise that will be injected into the IC, based on the knowledge of functional circuit blocks that are employed in the architecture.

The above ability is becoming crucial as systems are getting more broadband in nature, a rather extreme example being Ultra-Wideband. This is a newly emerging system which allows for high-speed, short-distance wireless communications by using a very large bandwidth for transmission and reception. The system uses a bandwidth from approximately 3.1 to 10.6GHz for UWB communications and requires a significant gate count to process the signal at baseband (the baseband being approximately 300MHz wide). In fact an OFDM based proposal for the standard has a large FFT engine that requires approximately seventy thousand gates clocking at nearly 500MHz.

With such high speeds of operation and gate densities and the wide input bandwidth of the system, it can be expected that these systems will be extremely sensitive to self-induced noise. The integrated thermal noise power in the UWB band is nearly -75dBm, which corresponds to a root-mean squared signal level of 40μV in a 50Ω load at the input of the receiver. Consider a digital core with approximately a hundred thousand gates, with a typical gate-to-substrate capacitance of 5fF per gate. This implies a total gate-to-substrate capacitance of 0.5nF. For a supply voltage of 1V, and a clocking frequency of 500MHz, the total power injected into the substrate is of the order of 250mW, if all gates switch simultaneously. Assuming an activity factor of 10%, the noise power injected into the substrate is 25mW, or 14dBm. The power within the band of interest will be approximately -13dB below this level, assuming a square-wave injection at 500MHz, with a $sinc^2$ roll off of the higher harmonics. If the system sensitivity is required to be thermal-noise limited, the coupled noise needs to be at least 10dB below the thermal

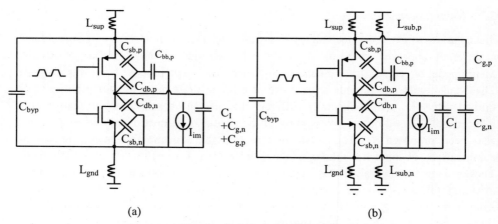

(a)   (b)

Fig. 6. Local and dedicated substrate ground configurations

noise floor. The required isolation through the substrate is thus of the order of 85dB. This is a very high level of isolation. It is likely, that these systems in their fully integrated forms will not be thermal noise limited, but limited by self-induced noise.

The potentially large contribution of self-induced noise to system sensitivity in systems such as the above leads us to explore faster and more efficient methods for noise estimation. In addition to the total power of self-induced noise, the ability to predict the temporal and spectral nature and the impact of this noise source on the link-budget during system and architecture design is very desirable.

We next propose a technique that has the potential to provide the above information using a functional level model of the circuit. This technique expresses the self-induced noise of particular functional blocks that are used in the design of the IC in terms of their functional parameters. This is in contrast to current macromodeling approaches that treat digital logic at a lower level of abstraction, and estimate the level of injected noise post-design. The utility of this technique stems from the observation that many communication transceivers utilize very similar functional cores such as analog-to-digital converters, Fast-Fourier Transform (and the Inverse FFT) cores, convolutional encoders and decoders and PLLs with various types of dividers, which are all described by a small set of parameters. For example, a flash ADC is well described by its resolution, the sampling rate, and the reference level. Further, the thermometer-to-binary encoders are well-known combinational logic blocks consisting of inverters, AND, and OR gates. The digital section of a flash ADC converter consists of an encoder which takes the thermometer code from the comparators and generates a binary equivalent output. The first level of this encoder consists of a set of inverters and AND gates that differentiate the input thermometer code and find the level at which the signal is located relative to the reference. This level is encoded as a binary equivalent through a combinational logic which we assume here to be composed of OR gates.

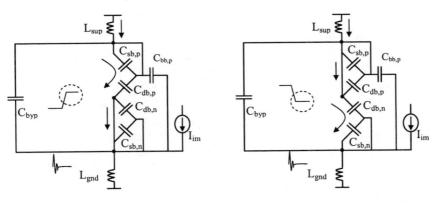

Fig. 7. Current flow for low-to-high and high-to-low transitions with local substrate ties

3.4. *Assumptions regarding switching mechanisms and physical media*

A signal change at the input in a circuit such as the above can be mapped deterministically to the number of gates that switch or change logic states. This is similar to estimating the dependence of quantization noise in a linear quantizer[31]. In order to facilitate the following analysis, we make certain assumptions regarding the type of switching currents and the substrate contacts on-chip.

In general, the substrate contacts can be implemented as local contacts to digital grounds or else by the use of dedicated substrate pins. We will assume that both NMOS and PMOS devices in the logic either use local grounds and supplies or dedicated pins. The two cases are depicted in Fig. 6 using simple static CMOS inverters. In each case we also assume the presence of a large on-chip bypass capacitor between the supply and ground points[10,11].

For each type of connection, certain observations can be made regarding the nature of current flow into the substrate node. With a local ground and supplies used as substrate contacts (Fig. 6a), the direction of current flow into the substrate/ground contact is independent of the direction of the input logic transition (Fig. 7). The current injection into the ground/substrate node itself consists of three dominant parts, the first is the current required to charge the net capacitance at the output node high or low, depending on the transition. For an output transition from low-to-high, the drain-to-body capacitance ($C_{db,n}$) of the NMOS devices is charged through the PMOS device which is on during this transition. Note that the PMOS drain-to-body capacitance ($C_{db,p}$) is discharged through the PMOS device itself, and does not contribute to current flow through the supply or ground. During a high-to-low transition, the PMOS drain-to-body capacitance ($C_{db,p}$) is charged through the NMOS device, while the NMOS drain-to-body capacitance ($C_{db,n}$) is locally discharged through the NMOS device. The source-to-body capacitances ($C_{sb,p}$ and $C_{sb,n}$) of both types of devices are effectively shorted by the local ground ties and to the first order do not contribute to current flow. Thus for local substrate-to-ground tie, the direction of current flow into the ground pin is identical for both transitions at the output.

The n-well-to-substrate capacitance contributed by the PMOS devices ($C_{bb,p}$) is effectively in parallel with the bypass capacitor. The second type of current flow occurs during the brief duration when both the PMOS and NMOS devices are on simultaneously. This represents a resistive connection between the supplies and the current flow here too is independent of the nature of output transition. Finally, impact ionization in the NMOS devices (I_{im}) also contributes to current flow through the local ground tie. As pointed out earlier, this current is physically unidirectional.

When dedicated substrate pins are used instead of local substrate connections to ground and supply (for NMOS and PMOS devices respectively), the nature of current flow into the substrate pins is substantially different from the above case. The first difference is that the source-to-body capacitors for both NMOS and PMOS devices are not locally shorted. These capacitors provide finite impedance between the ground/supply to substrate connections. The second difference is that the current through the drain-to-body capacitances is directional in this case. A high-to-low or low-to-high transition induces currents that flow in opposite directions. Due to this reason, some of the currents introduced into the substrate nodes are compensated by transitions with opposite polarity. It should be noted that unidirectional current flows such as in Fig. 6a, depend only on the total number of gate transitions. On the other hand for the substrate connection of Fig. 6b, the net ground bounce is induced by unidirectional currents related to the total number of gate transitions and the uncompensated transitions resulting from capacitive currents in the drain-to-body junctions.

Another assumption that we employ is that the digital core is sufficiently distant from the sensitive analog section, so that a combined aggregate impact at the local or dedicated substrate contacts is a good measure of the level of ground bounce introduced by the core into the substrate. We also assume that the inputs to the analog core are referred to an external ground, due to which the induced signal at the substrate contacts in the digital core is effectively the substrate noise as measured at the analog end.

3.5. *Switching activity in a thermometer to binary encoder*

If a flash ADC has a "b" bit resolution, a binary code is assigned to the input signal according to its state in the normalized range of $[-1{:}1/k{:}1]$ where k is 2^b. We have assumed the ADC reference voltage to be unity ($V_{ref}=1$) for simplicity. Therefore, the combinational logic block can have k different states for each bit resolution b. We will introduce below a technique for estimating the total ground bounce generated by the logic in the encoder, assuming a standard binary encoder of the type shown in Fig. 8. Ground-bounce is induced in the substrate any time a change at the input causes a state change in part or all of the encoder logic. The amount of ground-bounce depends on:

a)   Technology parameters, such as device junction and gate capacitance;
b)   Package parasitics, primarily the inductance in the ground and supply path;
c)   Structure of the encoder;
d)   The type of substrate connection (Fig. 6a/6b).

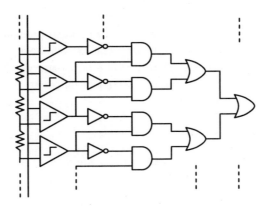

Fig. 8. Thermometer-to-binary encoder logic

We will assume here that the logic gates have their substrates connected to the local ground as in Fig. 6a. Thus current injection into the substrate is unidirectional and the amount of ground bounce is proportional to the total number of logic transitions for a given pair of inputs as pointed out earlier. Since the encoder state is deterministic, the total number of transitions for a change in the input can be determined exactly. In fact, for an input transition from $x(t)$ to $x(t+T_s)$, where T_s is the sampling period of the ADC, a transition function $T(x(t), x(t+T_s))$ can be defined. For each input transition therefore, the current step that will be induced will be scaled by $T(x(t),x(t+T_s))$. For a given input $x(t)$ therefore, the total ground-bounce will be proportional to a current given by:

$$I(t) = \sum_k T_{x_k, x_{k+1}} i(t - (k+1)T_s) .\qquad(12)$$

where $i(t)$ is the ground bounce current related to the switching of a single gate. The temporal characteristics of $i(t)$[11] depend on (a) and (b) above. Information regarding the value of capacitance switched at each node (gate capacitance and fan-out) can be included in T. If a unit gate size is assumed, T will relate only to the number of gate transitions. T is symmetric in its two variables, that is the total number of transitions are identical for input steps from x_1 to x_2 and x_2 to x_1.

It is easy to appreciate that $i(t)$ will be identical for a high-to-low and low-to-high transition only if the net capacitance to supply at a node equals the net capacitance to ground. Assuming a unit gate size for the encoder, where the net capacitance to supply at a unit gate does not equal the net capacitance to ground, the total current in (12) is given by

$$I(t) = \sum_k T_{0 \to 1}[x_k, x_{k+1}]i_{0 \to 1}(t - (k+1)T_s) + T_{1 \to 0}[x_k, x_{k+1}]i_{1 \to 0}(t - (k+1)T_s)] .\qquad(13)$$

where $T_{0 \to 1} / T_{1 \to 0}$ are the total number of levels undergoing a low-to-high/high-to-low transitions. These functions are not necessarily symmetric in their arguments. Current waveforms $i_{0 \to 1} / i_{1 \to 0}$ represent the unit gate impulses into the ground path arising from similar transitions.

While it is possible to use (13) above to compute $I(t)$, a more elegant approach is to express $T_{0 \to 1}$ and $T_{1 \to 0}$ in terms of the total transitions and the number of *uncompensated* gate transitions. For this purpose we define a function $U(x_1, x_2)$, which is the difference in the number of high levels in the logic between states x_1 and x_2. $T_{0 \to 1}$ is then given by *(T+U)/2*. Similarly $T_{1 \to 0}$ can be expressed as *(T-U)/2*. Using these functions in Eq. (13) we get:

$$I(t) = \frac{1}{2} \sum_k \left[(T_{x_k, x_{k+1}} + U_{x_k, x_{k+1}}) i_{0 \to 1}(t - (k+1)T_s) + (T_{x_k, x_{k+1}} - U_{x_k, x_{k+1}}) i_{1 \to 0}(t - (k+1)T_s) \right] \quad (14)$$

Eq. (14) allows for a more mathematically concise and elegant representation compared to Eq. (13). It is easier to determine T and U above for a given core because of their well-defined symmetry with respect to their inputs. For an input signal with known spectral characteristics, by using Eq. (14) we can determine the spectral characteristics of the ground bounce. We will illustrate this below by using our earlier example of a simple thermometer-to-binary encoder. For the introductory treatment here, we will assume that each node within the logic has identical unit capacitance that is equally divided between the supply and the ground. Thus T provides the total number of transitions for a change in the input between sampling instants kT_s and $(k+1)T_s$. T can be shown to be well approximated by the following function for any pair of inputs x_k and x_{k+1}.

$$T_{x_k, x_{k+1}} = \frac{2^{b+1} \left\{ 1 + \frac{|(x_k + x_{k+1})|^2}{6} \right\} |x_k - x_{k+1}|}{1 + e^{\frac{|(x_k + x_{k+1})|}{6}} |x_k - x_{k+1}|}. \quad (15)$$

Comparisons of this function with computed values of transitions are shown in Fig. 9 for two values of $(x_k + x_{k+1})$. The function is symmetric in x_k and x_{k+1}. Using Eq. (12), we see that the ground bounce caused by a transition from x_k to x_{k+1} is identical to that caused by the reverse transition. In the general case, for unequal supply and ground path capacitors, Eq. (14) will apply, and $I(t)$ will be depend on the sequence of the transition. For small values of x_K and x_{K+1},

$$T_{x_k, x_{k+1}} \cong 2^{b+1} |x_k - x_{k+1}|. \quad (16)$$

For a sinusoidal input, the input levels are given by:

$$x_k = aSin(\omega k T_s + \theta); x_{k+1} = aSin(\omega(k+1)T_s + \theta). \quad (17)$$

where a is the amplitude of the input sinusoid and θ is the initial phase, assumed random, since it is not synchronized with the sampler in general. The total number of logic transitions from Eq. (16) is then given by:

$$T_{x_k, x_{k+1}} \cong 2^{b+1} a |Sin(\omega T_s)| |Sin(\omega(k+1/2)T_s + \theta)|. \quad (18)$$

The above function can be applied in Eq. (12) to determine the total ground-bounce current. The fundamental frequency of the ground-bounce current will be 2ω in the above case. Further, for values of ω small compared to $1/T_s$, the amplitude of the ground-bounce

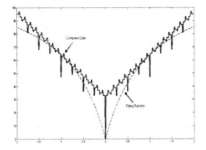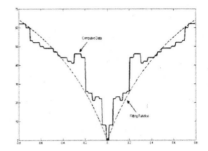

Fig. 9. Computed and modeled switching transitions for a 6 bit encoder for different values of $x_k + x_{k+1}$

currently will scale linearly with the input frequency. A band-limited white noise source will also show a similar frequency doubling and a scaling of the ground-bounce amplitude with frequency. It should be noted that this frequency dependence is not related to the properties of the substrate or package impedance, but is caused solely by the characteristics of the logic. Information regarding the substrate and package impedance will be implicitly contained in the unit current $i(t)$.

Eq. (18) can easily be extended to the case where the node capacitors to supply and ground are not identical. In fact, in that case, the ground bounce induced by a sinusoidal input of frequency ω can be shown to contain the fundamental frequency, unlike the above case.

The above analysis relates ground bounce to the total number of transitions and the number of uncompensated transitions, and for each case, maps the number of transitions on to parameters of the input waveform and the ADC encoder. It thus provides a convenient and straightforward estimate of the characteristics of the ground bounce, in relation to the parameters of the particular mixed-signal functional core. We believe that this technique can be generalized for several standard cores that are routinely employed in wireless transceivers. A library of functional, parameterized substrate noise signatures for several such mixed-signal cores should be feasible. It can be appreciated that such an approach will be greatly more useful and informative in comparison to an exact substrate noise estimation using full-extraction or even using gate-level noise signatures.

4. Conclusions

An overview of techniques for estimating and modeling self-induced noise related to substrate coupling and ground-bounce was provided. Full-extraction techniques based on the boundary element method and a fully numerical approach were outlined. The limitations of such approaches were discussed. Gate level macromodeling of substrate noise signatures and the considerable improvement allowed by the use of such techniques was discussed. Finally, a technique based on functional level parametric modeling of mixed-signal cores was presented using an example of a thermometer-to-binary encoder.

Such approaches can provide considerable information to the system architect without the need for complicated noise analysis. This information will be especially valuable in systems, where the sensitivity is either determined or otherwise greatly influenced by self-induced noise sources.

References

1. G.E. Moore, No exponential is forever: but "Forever" can be delayed!, *Proceedings of the International Solid State Circuits Conference*, 20 – 23 (2003).
2. T.A. Johnson, R.W. Knepper, V. Marcello and W. Wang, Chip substrate resistance modeling technique for integrated circuit design, *IEEE Transactions on Computer Aided Design*, **3**(4), 126-134 (1984).
3. T. Gabara, Reduced ground bounce and improved latch-up suppression through substrate conduction, *IEEE Journal of Solid State Circuits*, **23** (5), 1224-1232 (1988).
4. P.R. Gray and R.G. Meyer, Future directions in silicon ICs for RF personal communications, *Proceedings of the IEEE Custom Integrated Circuits Conference*, 83 – 90 (1995).
5. D.K. Su, M.J. Loinaz, S. Masui and B.A. Wooley, Experimental results and modeling techniques for substrate noise in mixed-signal integrated circuits, *IEEE Journal of Solid State Circuits*, **28**(4), 420-430 (1993).
6. R. Gharpurey, Modeling and analysis of substrate coupling in ICs, Ph. D. Dissertation, University of California at Berkeley, (1995).
7. E. Charbon, R. Gharpurey, P. Miliozzi, R.G. Meyer and A. Sangiovanni-Vincentelli, *Substrate Noise*, (Norwell, MA: Kluwer, 2001).
8. C. Hu, VLSI Electronics: Microstructure Science, Vol. 18, Academic Press, New York, 1981.
9. J. Briaire and K. S. Krisch, Substrate injection and crosstalk in CMOS circuits, *Proceedings of the IEEE Custom Integrated Circuits Conference*, 483-486 (1999).
10. M. Ingels and M.S.J. Steyaert , Design strategies and decoupling techniques for reducing the effects of electrical interference in mixed-mode IC's, *IEEE Journal of Solid-State Circuits*, 1136 – 1141 (1997).
11. P. Larsson, Resonance and damping in CMOS circuits with on-chip decoupling capacitance, *IEEE Transactions on Circuits and Systems—I*, 849-858 (1998).
12. N.K. Verghese, D.J. Allstot and S. Masui, Rapid simulation of substrate coupling effects in mixed-mode ICs, *Proceedings of the IEEE Custom Integrated Circuits Conference*, 1831-1834 (1993).
13. B.R. Stanisic, N.K. Verghese, R.A. Rutenbar, L.R. Carley and D.J. Allstot, Addressing substrate coupling in mixed-mode IC's: Simulation and Power Distribution Synthesis, *IEEE Journal of Solid State Circuits*, **29**(3), 226-238 (1994).
14. F. Clement, E. Zysman, M. Kayal and M. Declercq, LAYIN: Toward a global solution for parasitic coupling modeling and visualization, *Proceedings of the IEEE Custom Integrated Circuits Conference*, 537-540 (1994).
15. I. Wemple and A.T. Yang, Integrated circuit substrate coupling models based on Voronoi tessellation, *IEEE Transactions on Computer Aided Design*, **14**(12),1459-1469, (1995).
16. T. Smedes, N.P. van der Meijs and A.J. van Genderen, Boundary element methods for 3D capacitance and substrate resistance calculations in inhomogenous media in a VLSI layout verification package, Adv. in Engineering Software, **20**(1), 19-27 (1994).
17. R. Gharpurey and R.G. Meyer, Modeling and analysis of substrate coupling in integrated circuits, *IEEE Journal of Solid State Circuits*, **31**(3), 344-353 (1996).
18. J. Zhao, W. Dai, R.C. Frye and K.L. Tai, Green function via moment matching for rapid and accurate substrate parasitics evaluation, *Proceedings of the IEEE Custom Integrated Circuits Conference*, 371-374 (1997).

19. A.M. Niknejad, R. Gharpurey and R.G. Meyer, Numerically stable green function for modeling and analysis of substrate coupling in integrated circuits, *IEEE Transactions on Computer Aided Design*, **17**(4), 305-315 (1998).

20. J. P. Costa, M. Chou and L. M. Silveira, Efficient techniques for accurate modeling and simulation of substrate coupling in mixed-signal IC's, *IEEE Transactions on Computer Aided Design*, **18**(5), 587-607 (1999).

21. J. Kanapka and J. White, Highly accurate fast methods for extraction and sparsification of substrate coupling based on low-rank approximation, *Proceedings of the International Conference on Computer Aided Design*, 417-423 (2001).

22. M. Chou and J. White, Multilevel integral equation methods for the extraction of substrate coupling parameters in mixed-signal IC's, *Proceedings of the Design Automation Conference*, 20-25 (1998).

23. E. Schrik and N.P. van der Meijs, Combined BEM/FEM substrate resistance modeling, *Proceedings of the Design Automation Conference*, 771-776 (2002).

24. K. Joardar, A Simple approach to modeling cross- talk in integrated circuits, *IEEE Journal of Solid State Circuits*, **29**(10), 1212-1219 (1994).

25. A. Samavedam, A. Sadate, K. Mayaram and T. Fiez, A scalable substrate noise coupling model for design of mixed-signal IC's, *IEEE Journal of Solid State Circuits*, **35**(6), 895-904 (2000).

26. P. Miliozzi, L. Carloni, E. Charbon and A. Sangiovanni- Vincentelli, SubWave: a methodology for modeling digital substrate noise injection in mixed-signal ICs, *Proceedings of the IEEE Custom Integrated Circuits Conference*, 385-388 (1996).

27. E. Charbon, P. Miliozzi, L. Carloni and A. Ferrari and A. Sangiovanni-Vincentelli, Modeling digital substrate noise injection in mixed-signal IC's, *IEEE Transactions on Computer Aided Design*, **18**(3), 301-310 (1999).

28. A. Demir and P. Feldmann, Modeling and simulation of the interference due to digital switching in mixed- signal ICs, *Proceedings of the International Conference on Computer Aided Design*, 70-74 (1999).

29. R. C. Frye, Integration and electrical isolation in CMOS mixed-signal wireless chips, *Proceedings of the IEEE*, 444 – 455 (2001).

30. M. van Heijningen, J.Compiet, P. Wambacq, S. Donnay, M.E. Engels and I. Bolsens, Analysis and experimental verification of digital substrate noise generation for epi- type substrates, *IEEE Journal of Solid State Circuits*, **35**(7), 1002-1008 (2000).

31. R. M. Gray, Quantization noise spectra, *IEEE Transactions on Information Theory*, 1220-1244 (1990).

International Journal of High Speed Electronics and Systems
Vol. 15, No. 2 (2005) 297–317
© World Scientific Publishing Company

HIGH-SPEED OVERSAMPLING ANALOG-TO-DIGITAL CONVERTERS

AHMED GHARBIYA, TREVOR C. CALDWELL, AND D. A. JOHNS

Department of Electrical and Computer Engineering, University of Toronto
10 King's College Rd., Toronto, Ontario, CANADA, M5S 3G4

This paper is mainly tutorial in nature and discusses architectures for oversampling converters with a particular emphasis on those which are well suited for high frequency input signal bandwidths. The first part of the paper looks at various architectures for discrete-time modulators and looks at their performance when attempting high speed operation. The second part of this paper presents some recent advancements in time-interleaved oversampling converters. The next section describes the design and challenges in continuous-time modulators. Finally, conclusions are made and a brief summary of the recent state of the art of high-speed converters is presented.

Keywords: oversampling, delta-sigma, analog-to-digital

1. Introduction

Data conversion is an important operation that finds applications in many circuits today. *Delta-sigma* ($\Delta\Sigma$) modulation is a relative simple and low cost means of performing data conversion. While $\Delta\Sigma$ modulators can obtain a high dynamic range and excellent linearity with the use of a 1-bit quantizer, they are most often found in low-frequency applications since they oversample the data to achieve a high *signal-to-noise ratio* (SNR), thus limiting the input bandwidth by the speed at which the sampler can operate.

The sampler in a $\Delta\Sigma$ modulator must operate at a speed much greater than the bandwidth of the input signal since it must oversample the data. When standard CMOS technology is used, the sampling frequency of the modulator is limited to a few hundred megahertz. This limits the bandwidth of the input signal to around ten megahertz, depending on the *oversampling ratio* (OSR). Some methods of overcoming this bandwidth limitation include a feedforward architecture, time-interleaving discrete-time $\Delta\Sigma$ modulators, or using continuous-time circuitry.

This paper is laid out as follows: Section 2 discusses discrete-time $\Delta\Sigma$ modulator topologies; Section 3 demonstrates various time-interleaved discrete-time $\Delta\Sigma$ modulator topologies; Section 4 explains some design issues for continuous-time $\Delta\Sigma$ modulators; and Section 5 presents recent publications on high-speed $\Delta\Sigma$ modulators and states the conclusions.

2.   Single-Loop ΔΣ Modulator Topologies

This section provides a comparison of single-loop ΔΣ modulator topologies used for analog-to-digital converter applications that are suitable for high-speed implementation in deep sub-micron CMOS processes. To keep the scope of the analysis focused, the discussion is limited to ΔΣ modulators with pure differentiator type NTFs that employ internal quantizers with a sufficient number of levels to keep the modulator stable for any out-of-band gain. The relation of the topologies to their integrated-circuit implementation is emphasized.

The two main ΔΣ modulator topologies are the *chain of integrators with distributed feedback* (CIFB) and the *chain of integrators with weighted feedforward summation* (CIFF). To alleviate some of the drawbacks of the CIFB and CIFF topologies, the input-signal feedforward approach can be used as a modification. The resulting topologies, named *CIFB with input-signal feedforward* (CIFB-IF) and *CIFF with input-signal feedforward* (CIFF-IF), are discussed later. Note that, for the sake of simplicity, only modulators with all their zeros at dc will be discussed although these modulators can be modified with localized feedback to create non-dc zeros to optimize the NTF for a given OSR.

2.1. Chain of Integrators with Distributed Feedback

The simplest method to construct high order ΔΣ modulators is to cascade several integrators in the forward path, with each integrator receiving feedback from the quantizer to ensure stability. This CIFB topology is illustrated for a second-order modulator in Fig. 1.

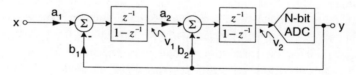

Fig. 1: Second-order CIFB modulator

Analysis of the linearized system with $a_1 = a_2 = b_1 = 1, b_2 = 2$ leads to the following results:

$$STF = \frac{y}{x} = z^{-2} \tag{1}$$

$$NTF = \frac{y}{q} = \left(1 - z^{-1}\right)^2 \tag{2}$$

$$v_1 = z^{-1}\left(1 + z^{-1}\right)x - z^{-1}\left(1 - z^{-1}\right)q \tag{3}$$

$$v_2 = z^{-2}x - z^{-1}\left(2 - z^{-1}\right)q \tag{4}$$

where q is the quantization noise from the ADC, and v_1 and v_2 are the signals at the outputs of the first and second integrators, respectively. The STF exhibits an all-pass response and the NTF provides a second-order pure differentiator type high-pass response.

The main advantages of the CIFB topology are that it is easy to implement with low sensitivity to component variations. The main disadvantage of this topology is that the signals at the output of the integrators are a function of the input-signal as given in Eqs. (3) and (4) above, resulting in two effects. First, the signal swing at the output of the opamps is large which makes their implementation in the low-voltage, deep sub-micron technology more difficult. Second, opamp nonlinearities generate harmonic distortion that depends on the input-signal. The opamp distortion can severely limit the achievable SQNR. Another disadvantage of the CIFB topology is that the NTF and STF cannot be set independently. Therefore, if we pick a certain NTF, then the STF is fixed.

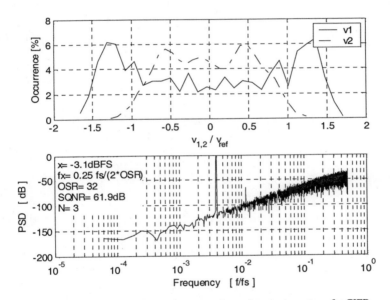

Fig. 2: Signal swing at Opamp outputs and sample output spectrum for CIFB

The CIFB topology is simulated using MATLAB® and Simulink®. The probability density function of integrator outputs and a sample output spectrum including opamp third-order distortion are shown in Fig. 2. The third-order distortion model is in the form of a power series with the third-order term corresponding to 1% third-order harmonic distortion for full scale signal. Simulations indicate the signal swings can be more than 1.5 times larger than the ADC reference voltage. On the other hand, the input-signal range is from 50 to 80% of the ADC reference voltage and depends on the loop order and number of bits in the quantizer.[1] Therefore, the input-signal is going to be relatively small when compared to other topologies, and to meet thermal noise requirements the capacitor sizes must be larger, leading to greater power dissipation. The third harmonic generated by the opamp nonlinearity is clear in the output spectrum shown in Fig. 2. Distortion severely reduces the SQNR of the CIFB topology from the ideal 76dB to 62dB for the example shown in Fig. 2.

The CIFB is the most commonly used topology to implement ΔΣ modulators. An example of the CIFB topology is implemented as a third-order CIFB ΔΣ modulator using a 4-bit internal quantizer and operating with a sampling frequency of 100MHz.[2] The modulator achieves an SNDR of 67dB and a peak SNR of 68dB with a 12.5MS/s conversion rate. The modulator is implemented in 0.65μm technology and powered with 5V supply while consuming 380mW.

2.2. Chain of Integrators with Weighted Feedforward Summation

Distributed feedback was used to ensure stability of the cascade of integrators in the forward path. Alternatively, weighted feedforward paths can be used to establish stability. The resulting chain of integrators with the CIFF topology for a second-order modulator is shown in Fig. 3.

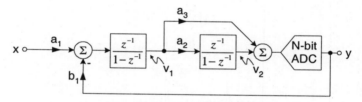

Fig. 3: Second-order CIFF modulator

Analysis of the linearized system with $a_1 = a_2 = b_1 = 1, a_3 = 2$ leads to the following results:

$$STF = \frac{y}{x} = z^{-1}(2 - z^{-1}) \tag{5}$$

$$NTF = \frac{y}{q} = (1 - z^{-1})^2 \tag{6}$$

$$v_1 = z^{-1}(1 - z^{-1})x - z^{-1}(1 - z^{-1})q \tag{7}$$

$$v_2 = z^{-2}x - z^{-2}q \tag{8}$$

where q is the quantization noise from the ADC, and v_1 and v_2 are the signals at the outputs of the first and second integrators, respectively.

The CIFF improves the performance of CIFB in terms of the signals at the output of the integrators. As can be seen from Eq. (7), the signal at the output of the first opamp contains a first-order noise shaped input-signal component in addition to shaped quantization noise. This reduces signal swing and reduces dependence of the distortion on the input-signal. Both of these benefits are illustrated in Fig. 4. The signal swing at the output of the first opamp is significantly reduced and the output spectrum does not show harmonic distortion. The second opamp still contains an input-signal component, however, nonlinearities at this stage are not as important since they are second-order noise shaped when referred back to the input.

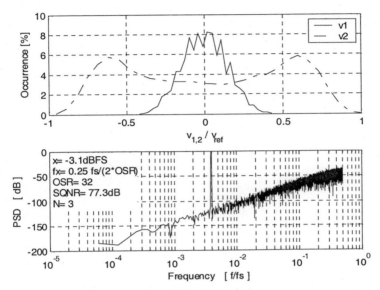

Fig. 4: Signal swing at Opamp outputs and sample output spectrum for CIFF

The main disadvantage of the CIFF topology can be seen by investigating its STF given in Eq. (5). The STF has a high frequency boost with a gain of one at low frequencies and three at high frequencies. The amplification of the out-of-band frequencies due to the high frequency boost can overload the quantizer and drive the modulator into instability. Unfortunately, the NTF and STF are not independent; therefore, the high frequency boost in STF is fixed by the choice of the NTF.

One of the fastest CMOS $\Delta\Sigma$ modulators reported in literature is implemented using the CIFF topology where a fifth-order CIFF $\Delta\Sigma$ modulator uses a 4-bit internal quantizer and operates at a 200MHz sampling frequency.[3] The modulator achieves an SNDR of 72dB with a peak SNR of 82dB at a conversion rate of 25MS/s. This performance is achieved in 0.18μm CMOS technology.

2.3. CIFB with Input-Signal Feedforward

The input-signal component at opamp outputs in the CIFB topology can be eliminated by feeding the input-signal forward such that the input-signal components cancel out. The resulting CIFB-IF topology is illustrated for a second-order modulator in Fig. 5.

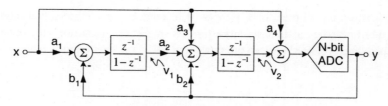

Fig. 5: Second-order CIFB-IF modulator

Analysis of the linearized system with $a_1 = a_2 = a_4 = b_1 = 1, a_3 = b_2 = 2$ leads to the following results:

$$STF = \frac{y}{x} = 1 \tag{9}$$

$$NTF = \frac{y}{q} = \left(1 - z^{-1}\right)^2 \tag{10}$$

$$v_1 = -z^{-1}\left(1 - z^{-1}\right)q \tag{11}$$

$$v_2 = -z^{-1}\left(2 - z^{-1}\right)q \tag{12}$$

where q is the quantization noise from the ADC, and v_1 and v_2 are the signals at the outputs of the first and second integrators, respectively.

The input-signal feedforward modifies v_1, v_2, and the STF without affecting the NTF. The signals v_1 and v_2 are free of the input-signal component. Therefore, the signal swings are smaller and the distortion generated by the opamps is input-signal independent. These advantages are illustrated in Fig. 6.

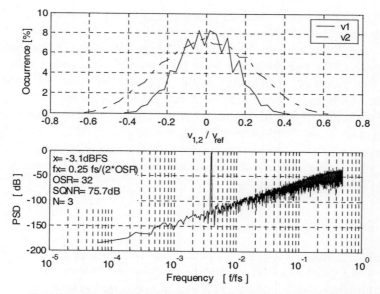

Fig. 6: Signal swing at Opamp outputs and sample output spectrum for CIFB-IF

The disadvantage of the CIFB-IF topology is the increased loading that the input has to drive, which can be particularly large for higher order modulators. This is because of the distributed feedforward paths that are needed to achieve the input-signal cancellation. In the second-order case, for example, there is the main sampling capacitor at the input as well as two extra sampling capacitors to feed the input-signal forward. It should be mentioned that the extra capacitors are usually smaller than the input sampling capacitor because the thermal noise on these capacitors is noise shaped and therefore, their size can be smaller.

An example of the CIFB-IF topology is implemented as a second-order modulator using a single-bit internal quantizer and operating with a sampling frequency of 105MHz.[4] The modulator achieves a dynamic range of 88dB with a peak SNR of 82dB. The modulator is implemented in 0.13μm CMOS technology and powered with a 1.5V supply while consuming only 8mW of power.

2.4. CIFF with Input-Signal Feedforward

The main problem with the CIFF topology is the high frequency boost in the STF. The input-signal feedforward concept can be used to modify the STF of the CIFF topology without affecting the NTF. The CIFF-IF topology is illustrated in Fig. 7 for a second-order modulator.[5]

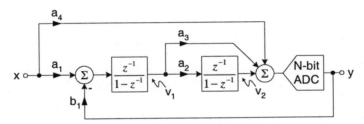

Fig. 7: Second-order CIFF-IF modulator

Analysis of the linearized system with $a_1 = a_2 = a_4 = b_1 = 1, a_3 = 2$ leads to the following results:

$$STF = \frac{y}{x} = 1 \qquad (13)$$

$$NTF = \frac{y}{q} = (1 - z^{-1})^2 \qquad (14)$$

$$v_1 = -z^{-1}(1 - z^{-1})q \qquad (15)$$

$$v_2 = -z^{-2}q \qquad (16)$$

where q is the quantization noise from the ADC, and v_1 and v_2 are the signals at the outputs of the first and second integrators respectively.

The input-signal feedforward changes the problematic high frequency boost in the STF of the CIFF topology to an all-pass STF in the CIFF-IF topology with no effect on the NTF. It is interesting to note that this modulator achieves the smallest signal swings at the output of the opamps among the topologies discussed, as seen in Fig. 8.

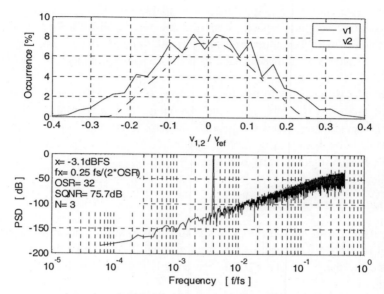

Fig. 8: Signal swing at Opamp outputs and sample output spectrum for CIFF-IF

3.   Time-Interleaved $\Delta\Sigma$ Modulators

This section presents time-interleaved $\Delta\Sigma$ modulator topologies based on block filtering theory. The usual system level design parameters for $\Delta\Sigma$ modulators are the loop-order, OSR, and the number of bits in the internal quantizer. High-speed applications require low OSRs, thereby, limiting the choices available for the designer. One method to add another degrees of freedom is to use parallel $\Delta\Sigma$ structures. The simplest method of making parallel converters is through the use of *time-interleaving* (TI) which is simply a time-division multiplexing scheme where an array of individual converters are clocked at different instants in time. Unfortunately, exploiting simple time-interleaved parallelism is not a straightforward process for $\Delta\Sigma$ modulators due to their recursive nature. Straightforward TI adaptation to $\Delta\Sigma$ modulators results in a 3dB improvement in the SNR for each doubling of converters regardless of the order of the modulator. To overcome this problem, different schemes of parallel modulators have been devised. They can be classified in one of three main categories: *frequency division multiplexing* (FDM),[6] *code division multiplexing* (CDM),[7] and *time division multiplexing* (TDM).[8]

TDM can be implemented by deploying the theory of block digital filtering. The principle of block digital filtering is based on transforming a linear time-invariant (LTI) single-input single-output system (SISO) with transfer function $H(z)$ to an equivalent multi-input multi-output system with transfer function $\overline{H}(z)$, as shown in Fig. 9.

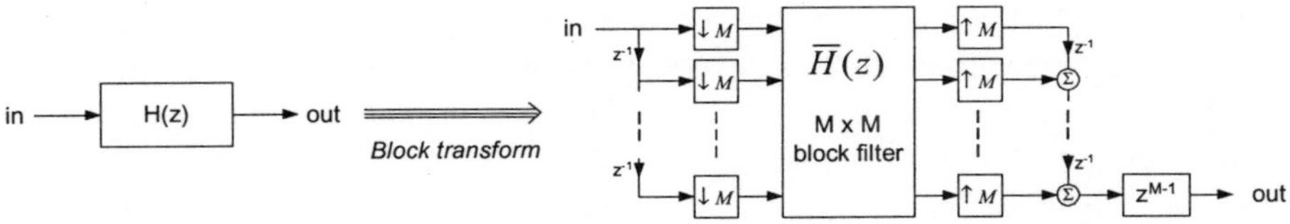

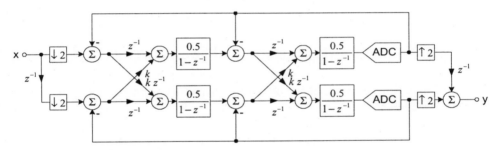

Fig. 9: H(z) and its blocked version with block length M

The internal circuitry of the block filter operates in parallel and at a reduced rate by the factor M. For example, using this transformation for a ΔΣ modulator with M=2 allows the internal modulators to either operate at half-speed for the same resolution, or at enhanced resolution for the same speed. This improvement is significant in wide bandwidth applications where the sampling speed is limited by the technology and resolution requirements.

The block digital filtering has facilitated the design and construction of a true TI ΔΣ modulator.[8] A second-order, time-interleaved by 2 (M=2), CIFB ΔΣ modulator is shown in Fig. 10 as an example of the technique.[8]

Fig. 10: Second-order time-interleaved by 2 CIFB ΔΣ modulator

The k-factor shown in Fig. 10 is used to deal with the issue of opamp DC offsets.[8] DC offsets are problematic in time-interleaved modulators because the difference in offset between the two branches drives the modulator to instability. Reducing the cross-coupling coefficients gives more control to each parallel ΔΣ modulator, thus enabling the negative feedback loop to adjust; which maintains DC stability. However, reducing k from unity modifies the STF and results in an increase of the quantization noise in the signal band, thereby reducing the SQNR. The choice of k is a tradeoff between the offset value that the modulator can tolerate and the achievable SNR. A time-interleaved modulator that does not suffer from DC offsets is presented below.

A high-speed input demux is needed at the input of the modulator to sample the input-signal and distribute it to the individual internal modulators. The demux operates at the full speed of the overall modulator. For example, the demux in a time-interleaved by 4 modulator operates at four times the speed of the individual ΔΣ modulators. The high-speed demux can become the limiting factor in the performance of the modulator especially for higher-order TI structures (M>2). A solution for the demux problem for M=2 is to sample each branch in the time-interleaved modulator at a different phase of the two non-overlapping phases.[8] Therefore, the demux is inherent in the operation of the modulator. Another more general solution that can be used for any M is called the zero-insertion interpolation technique,[9] which is shown in Fig. 11 for M=2 second-order CIFB topology. The *zero-insertion time-interleaved* (ZI-TI) modulator samples the input-signal at the operating frequency of the individual $\Delta\Sigma$ modulator and applies these samples to the first branch only with the inputs to the others grounded.

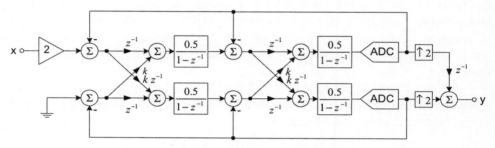

Fig. 11: Second-order ZI-TI with M=2 CIFB $\Delta\Sigma$ modulator

The sampled input must be amplified (by M) to compensate for the lost signal power resulting from supplying zero input instead of the input-signal to the other branches. The ZI-TI modulator still suffers from DC offsets and therefore the cross-coupling coefficient k must be set appropriately.

A new *modified time-interleaved* (MTI) $\Delta\Sigma$ modulator that eliminates the input demux and alleviates the DC offset problem is shown in Fig. 12. The modulator can be derived starting from the TI (Fig. 10) by removing the demux and applying the input-signal to both branches simultaneously. The resulting second-order MTI CIFB $\Delta\Sigma$ modulator, after some modifications, is shown in Fig. 12. The MTI uses two integrators only instead of the four used in TI, and in general, it uses the same number of integrators as the basic $\Delta\Sigma$ modulator for a given loop order. Therefore, there is no DC offset difference between the two branches that would otherwise lead to instability. In summary, the MTI eliminates the high-speed analog demux, alleviates the DC offset problems, and uses fewer integrators.

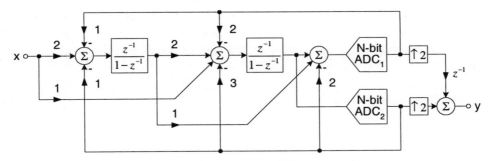

Fig. 12: Second-order modified time-interleaved by 2 CIFB $\Delta\Sigma$ modulator

Removing the demux at the input has some consequences. Analysis of the linearized system of Fig. 12 leads to the following results:

$$y = z^{-2}\left(1 + z^{-1}\right)x + z^{-1}\left(1 - z^{-1}\right)^2 q_1 + \left(1 - z^{-1}\right)^2 q_2 \qquad (17)$$

where q_1 and q_2 are the quantization noise from ADC_1 and ADC_2 respectively. Due to the output mux, the quantization noise q_1 is only added to the output once for every two samples, which is also true for q_2. Therefore, the overall noise contribution can be rewritten as:

$$NTF = \frac{y}{q} = \left(1 - z^{-1}\right)^2 \qquad (18)$$

which is simply second-order noise shaped. Clearly, the removal of the demux does not affect the TI NTF, however the STF is affected. The first term in the STF is z^{-2}, which is the expected STF of a second-order CIFB modulator. The second term, $\left(1 + z^{-1}\right)$, resulted from the removal of the input demux. The extra term adds a notch at half the sampling frequency and filters the amplitude response of the STF as shown in Fig. 13. Due to oversampling, the frequency variation is not significant within the signal band. Also, it can be easily compensated for in the digital domain.

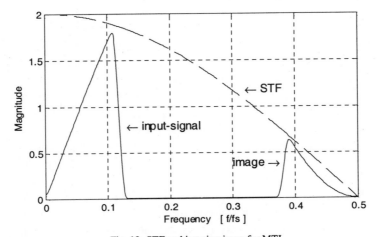

Fig. 13: STF and imaging issue for MTI

Another effect of removing the demux is that the signal is under the influence of the upsamplers only. The effect of upsampling by M is M-fold compression and repetition of the frequency-domain magnitude response.[10] The process generates images shaped by the STF at frequencies less than half the sampling frequency as shown in Fig. 13 for a sample input spectrum. The design of the anti-aliasing and decimation filters should take the imaging issue into account.

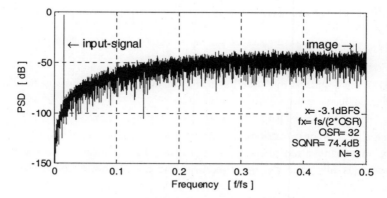

Fig. 14: Sample output spectrum for MTI

A sample output spectrum for the MTI is shown in Fig. 14. It also highlights the shaped image of the input-signal. For this example, the integrators are clocked at half the rate as the single-loop topologies presented earlier. The MTI modulator achieved a slightly lower SQNR because the STF attenuates the input-signal which reduces the input power.

4.   Continuous-Time $\Delta\Sigma$ Modulators

Employing continuous-time loop filters instead of discrete-time loop filters is one way to increase the input signal bandwidth. The main advantage of continuous-time filters is that no sampling is performed within the filters, so the restriction of the maximum sampling frequency is only imposed on the quantizer, as well as on the feedback DAC. Practically, continuous-time modulators can operate with clock frequencies about 2-4 times greater than regular discrete-time modulators, while suffering from reduced linearity and accuracy.[11] Also, continuous-time modulators eliminate the need for an anti-aliasing filter on the input since it is inherent in the signal transfer function (STF).

There do exist some disadvantages of continuous-time $\Delta\Sigma$ modulators when compared to discrete-time modulators. High-speed continuous-time modulators suffer more severely from two non-idealities, namely excess loop delay and DAC clock jitter. The following section will explain some of the more important design considerations, including a brief description on how to design a continuous-time $\Delta\Sigma$ modulator from a discrete-time ΔΣ modulator while avoiding excessive STF peaking, and some methods of reducing excess loop delay and DAC clock jitter.

4.1. Discrete-to-Continuous Transform

To design a continuous-time ΔΣ modulator, a discrete-time ΔΣ modulator may be designed and simulated, and then a conversion between the two modulators can be performed to realize the desired loop filters of the continuous-time ΔΣ modulator. One method of finding equivalence between a continuous-time and discrete-time modulator is to recognize that an implicit sampling occurs in the quantizer of the continuous-time modulator.[12] If the open-loop modulators are analyzed, as shown in Fig. 15, the two modulators are equivalent as long as the outputs are equal at the sampling instants. Therefore, if $w[n] = w(t)|_{t=nT}$ for all n, then the loop filters will be equivalent. The resulting condition for the two filters $B(z)$ and $B(s)$ to be equivalent is:[13]

$$Z^{-1}\{B(z)\} = L^{-1}\{R(s) \cdot B(s)\}|_{t=nT} \tag{19}$$

This transformation is known as the impulse-invariant transformation,[14] where Z^{-1} represents the inverse z-transform, L^{-1} represents the inverse Laplace transform, and $R(s)$ represents the DAC pulse.

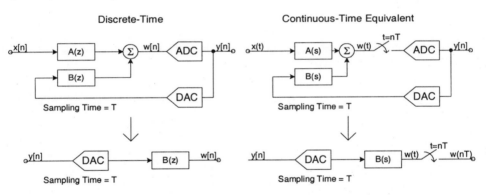

Fig. 15: Open loop continuous-time equivalent of discrete-time modulator

To properly account for the shape of the DAC pulse, Eq. (19) is rewritten with the DAC pulse $R(s)$ represented by:[12]

$$R(s) = \frac{e^{-\alpha s} - e^{-\beta s}}{sT} \tag{20}$$

The time domain representation of this DAC pulse transfer function $R(s)$ is:

$$r(t) = \begin{cases} 1, & \alpha \le t < \beta, \ 0 \le \alpha < \beta \le T \\ 0, & otherwise \end{cases} \tag{21}$$

Eqs. (20) and (21) assume that the pulse is rectangular and has a magnitude of one, lasting from $t = \alpha$ to $t = \beta$.

As an example, if a discrete-time $\Delta\Sigma$ modulator were designed with an NTF of $H(z) = (1 - z^{-1})^2$ (with an STF $G(z) = z^{-1}$), then the continuous-time $\Delta\Sigma$ modulator would be designed as follows:

1) Referring to Fig. 15, $A(z)$ and $B(z)$ are found as follows from the given NTF and STF:

$$B(z) = \frac{-2z^{-1} + z^{-2}}{1 - 2z^{-1} + z^{-2}} = \frac{-2z + 1}{z^2 - 2z + 1} \tag{22}$$

$$A(z) = \frac{z^{-1}}{1 - 2z^{-1} + z^{-2}} = \frac{z}{z^2 - 2z + 1} \tag{23}$$

2) The filters $A(z)$ and $B(z)$ are dissected into their partial fraction representation:

$$B(z) = \frac{-1}{z^2 - 2z + 1} + \frac{-2}{z - 1} \tag{24}$$

$$A(z) = \frac{1}{z^2 - 2z + 1} + \frac{1}{z - 1} \tag{25}$$

3) Using Eqs. (24) and (25) for $A(z)$ and $B(z)$ (where $R(s)$ would have $\alpha = 0$ and $\beta = T$), the resulting equivalent continuous-time filters are (using the transforms $\frac{1}{z - 1} \to \frac{1}{Ts}$, $\frac{1}{(1 - z)^2} \to \frac{-Ts + 2}{2T^2 s^2}$):

$$B(s) = \frac{Ts - 2}{2T^2 s^2} + \frac{-2}{Ts} = \frac{-3Ts - 2}{2T^2 s^2} \tag{26}$$

$$A(s) = \frac{-Ts + 2}{2T^2 s^2} + \frac{1}{Ts} = \frac{Ts + 2}{2T^2 s^2} \tag{27}$$

4) These loop filters $A(s)$ and $B(s)$ can be converted into a $\Delta\Sigma$ modulator topology. An example of one possible modulator is shown in Fig. 16.

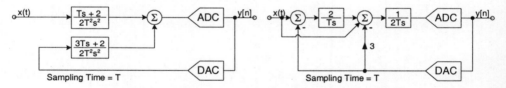

Fig. 16: Continuous-time modulator to realize derived loop filters

4.2. Signal Transfer Function

In the example above, $A(z)$ was designed to be $\dfrac{z}{z^2 - 2z + 1}$. An extra delay in this path does not change the discrete-time transfer function, but it can alter the continuous-time transfer function when the discrete-to-continuous transform is applied. The result is that there can be potentially more or less peaking in the STF. This can be hazardous since it can cause instability in the modulator for certain input frequencies. Therefore, the STF should be computed and analyzed to ensure minimal peaking.

Assuming a linearzied model of the $\Delta\Sigma$ modulator where the quantizer is replaced by a unity-gain block, the STF can be computed as the input loop filter $A(s)$ multiplied by the discrete-time NTF. The NTF is discrete-time due to the sampling in the quantizer. To make them both a function of the same frequency, the discrete-time transfer function is evaluated at $z = e^{j2\pi fT}$ while the continuous-time transfer function is evaluated at $s = j2\pi fT$.[11] When the resulting transfer function is plotted, the inherent anti-aliasing property of the continuous-time STF is apparent.

Using an $A(z)$ of higher order to illustrate the change in the STF, the functions $\dfrac{1}{z^3 - 3z^2 + 3z - 1}$, $\dfrac{z}{z^3 - 3z^2 + 3z - 1}$ and $\dfrac{z^2}{z^3 - 3z^2 + 3z - 1}$ can all be converted to their equivalent continuous-time input loop filters $A(s)$. These discrete-time filters $A(z)$ all result in the same transfer functions, except for a difference in the latency at the output. But when the equivalent $A(s)$ for all three cases is multiplied by the NTF $H(z) = (1 - z^{-1})^3$ (evaluated at $z = e^{j2\pi fT}$), the resulting continuous-time STFs are shown in Fig. 17. Only two graphs are plotted since the continuous-time STFs are the same when $A(z)$ equals either $\dfrac{1}{z^3 - 3z^2 + 3z - 1}$ or $\dfrac{z^2}{z^3 - 3z^2 + 3z - 1}$. But in both these cases, the STF peaks 1.4dB higher than when $A(z) = \dfrac{z}{z^3 - 3z^2 + 3z - 1}$, resulting in a potentially less stable $\Delta\Sigma$ modulator due to this unwanted gain.

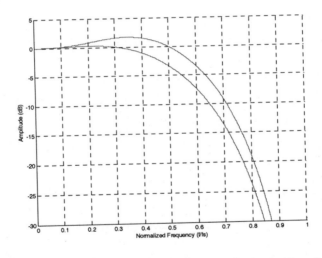

Fig. 17: Comparison of continuous-time STF for various discrete-time input transfer functions

4.3. Excess Loop Delay

One of the major difficulties with continuous-time $\Delta\Sigma$ modulators is that a small delay t_d exists between the quantizer sampling instant and when the DAC pulse is valid because the transistors cannot switch instantaneously. This is known as excess loop delay.[12] The excess loop delay in a continuous-time modulator effectively increases the order of the modulator if the pulse enters the next clock period, as demonstrated in Fig. 18. When the pulse enters the adjacent clock period, the resulting order of the equivalent discrete-time transfer function increases.[12] This increase in order can be illustrated by modeling the output of the DAC pulse with excess loop delay t_d as the sum of two pulses as follows:

$$DAC_{(td,T+td)}(t) = DAC_{(td,T)}(t) + DAC_{(0,td)}(t-T) \qquad (28)$$

where $DAC_{(td,1)}(t)$ represents a pulse from $\alpha = t_d$ to $\beta = T$, and $DAC_{(0,td)}(t-T)$ represents a pulse from $\alpha = 0$ to $\beta = t_d$ that has been shifted in time by the sampling period T. The resulting z-transform will be the sum of two pulses, one of which has an extra z^{-1} term due to the delayed DAC pulse $DAC_{(0,td)}(t-T)$, and this will contribute to the increased order of the transfer function. Using this analysis, the transfer function:[12]

$$H(z) = \frac{-2z+1}{(z-1)^2} \qquad (29)$$

becomes

$$H(z) = \frac{(-2 + 2.5t_d - 0.5t_d^2)z^2 + (1 - 4t_d + t_d^2)z + (1.5t_d - 0.5t_d^2)}{z(z-1)^2} \qquad (30)$$

where the extra order is evident. This reduces the stability of the $\Delta\Sigma$ modulator, and is more significant at larger values of the excess loop delay t_d. Also, the excess loop delay increases the noise floor of the $\Delta\Sigma$ modulator. This analysis facilitates simulation of the equivalent discrete-time modulator to fully evaluate the acceptable excess loop delay in a given $\Delta\Sigma$ modulator design.

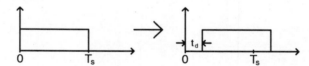

Fig. 18: Excess loop delay in a full period DAC pulse

The use of an additional feedback term directly into the quantizer from the DAC in the continuous-time $\Delta\Sigma$ modulator has been shown to mitigate the effects of excess loop delay.[15, 16] In fact, it was shown that in some cases a delay greater than T can be used, as long as the delay is properly chosen.[15] This technique is shown in Fig. 19 with the extra f path.

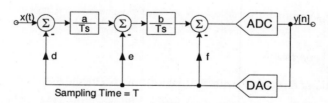

Fig. 19: Additional path to reduce effects of excess loop delay

Another simple way of reducing the effects of excess loop delay is to use *return-to-zero* (RZ) DAC pulses. Under this condition, the width of the pulse in Fig. 18 can be adjusted so that it does not extend beyond the sampling interval T. The value of the continuous-time filters must then be adjusted accordingly since the α and β parameters of Eqs. (20) and (21) will be modified. One difficulty with this solution is that the loop filters of the $\Delta\Sigma$ modulator will have been designed specifically for a given α and β, and changes in the value t_d will alter the rise and fall instants of the feedback pulse.

Therefore, these rise and fall instants should be very well controlled with respect to the quantizer clock, especially at high speeds. It is also possible to introduce a variable delay block between the quantizer clock and the RZ DAC pulse clock so that this delay value can be manually or automatically tuned for increased precision.

4.4. Clock Jitter

Clock jitter is statistical variations of clock edges.[17] Two clocks are present in a CT $\Delta\Sigma$ modulator and both can be affected by clock jitter. One of the clocks controls the decision instant of the quantizer (or comparator) while the other clock controls the DAC output. Since the output of the comparator is shaped by the NTF (like the quantization noise), the impact of this error will be relatively small. Conversely, the output of the DAC is shaped by the STF because this signal adds to the input signal and thus, the impact of this error affects the passband noise in the $\Delta\Sigma$ modulator.[18]

There are two varieties of clock jitter, delay clock jitter and pulse-width clock jitter. In a second-order $\Delta\Sigma$ modulator, the delay clock jitter is affected by the NTF while the pulse-width clock jitter manifests itself as white noise.[19] Thus, the pulse-width clock jitter degrades the SNR of the $\Delta\Sigma$ modulator more severely since the white noise fills in the notch in the signal band, directly reducing the noise floor in the band of interest. Therefore, the clock jitter discussed will be the pulse-width clock jitter incurred in the DAC.

Discrete-time $\Delta\Sigma$ modulators are relatively insensitive to pulse-width clock jitter since they utilize switched-capacitor circuits. The insensitivity is due to the sloping form of the feedback pulse.[17] Because most of the charge transfer in a switched-capacitor circuit occurs at the beginning of the clock period, clock jitter introduces a minimal amount of error in the charge lost ΔQ_D (see Fig. 20).[20] The capacitor is discharged over a switch with very low on-resistance, thus reducing the value of $\tau = RC$ and causing a fairly steep slope as the DAC discharges.[20] In contrast, continuous-time $\Delta\Sigma$ modulators transfer charge at a constant rate over the clock period (ideally), and thus, the charge loss ΔQ_C, due to a timing error, is proportionally much greater than that of the discrete-time $\Delta\Sigma$ modulator.

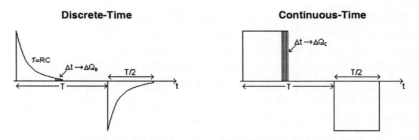

Fig. 20: Clock jitter in discrete-time and continuous-time modulators

The clock jitter with RZ DAC pulses is going to be more detrimental than that for the *non-return-to-zero* (NRZ) DAC pulses for a couple of reasons. First, with RZ DAC pulses, since the width of the pulse is smaller, this proportionally results in a larger percentage difference of the total integrated signal when jitter is introduced, when compared to a full period NRZ DAC pulse. Second, in every clock period, the RZ DAC pulse returns to zero, independent of the previous value, and thus introduces clock jitter in every pulse. Conversely, in the NRZ DAC pulse, adjacent pulses may have the same value, thus introducing no clock jitter in that period. And finally, similar to the last point, in a multi-bit $\Delta\Sigma$ modulator the output of adjacent DAC pulses typically do not span the entire range of the feedback DAC, and are typically a few levels apart. Therefore, the clock jitter will be present in a smaller fraction of the total pulse area. However, when a RZ DAC pulse is used, the output pulse must return to zero during every period, and thus, the change in level is greater than that of an NRZ pulse. On average, this increases the presence of clock jitter when using an RZ pulse as compared to an NRZ pulse in a multibit feedback DAC.

The effects of clock jitter for two low-pass $\Delta\Sigma$ modulators with single-bit quantizers were analyzed, where one used RZ DAC pulses, and the other used NRZ DAC pulses.[20] It was found that the noise power in the case with RZ DAC pulses was about 3 times worse than that with NRZ DAC pulses.

4.5. Time-Interleaving

Continuous-time $\Delta\Sigma$ modulators can also be time-interleaved. As an example, the second-order discrete-time modulator shown in Fig. 11 can be converted to its time-interleaved by 2 continuous-time equivalent with various manipulations of the loop filters (since there are now clearly more than just $A(z)$ and $B(z)$ as the loop filters). The resulting modulator is shown in Fig. 21 where the ADCs and DACs are operating at half of the original frequency, meaning that clock jitter has about half the impact on the $\Delta\Sigma$ modulator. Furthermore, the higher bandwidth requirements on an RZ DAC operating at high-speeds (assuming the choice of an RZ DAC pulse to reduce the excess loop delay) is reduced by a factor of two by time-interleaving.

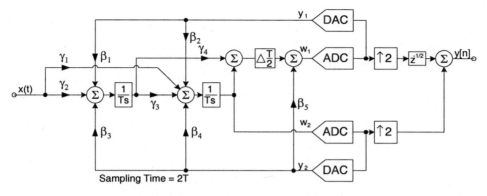

Fig. 21: Time-interleaved continuous-time $\Delta\Sigma$ modulator

5. Conclusions

This paper discussed a variety of techniques for designing high-speed oversampling analog-to-digital converters including an input feedforward architecture, time-interleaving and continuous-time modulators. Some design challenges were highlighted and potential solutions described. Some recent publications of higher-speed CMOS $\Delta\Sigma$ modulators are shown in Table 1 and give an indication of the present state-of-the-art for input signal bandwidths greater than 5MHz. While continuous-time modulators tend to dominate at higher input signal bandwidths, we see from Table 1 that excellent results can still be obtained using a discrete-time modulator through careful architecture and circuit design.[3]

Table 1: Recently published high-speed $\Delta\Sigma$ modulators

Ref.	Technology	Sampling Frequency	SNDR	Power	Bandwidth	Topology
2	0.65 µm CMOS	100 MHz	67 dB	295 mW	6.25 MHz	DT (CIFB)
21	0.13 µm CMOS	80 MHz	50 dB	80 mW	10 MHz	CT
22	0.13 µm CMOS	160 MHz	57 dB	122 mW	10 MHz	CT
3	0.18 µm CMOS	200 MHz	72 dB	200 mW	12.5 MHz	DT (CIFF)
23	0.13 µm CMOS	300 MHz	64 dB	70 mW	15 MHz	CT
21	0.13 µm CMOS	160 MHz	50 dB	120 mW	20 MHz	CT

References

1. S. R. Norsworthy, R. Schreier, and G. C. Temes, Delta-Sigma Data Converters: Theory, Design, and Simulation. Piscataway, NJ: IEEE Press, pp.157, 1997.

2.   Y. Geerts, M. S. J. Steyaert, W. Sansen, "A high-performance multibit $\Delta\Sigma$ CMOS ADC," *IEEE JSSC*, vol. 35, no. 12, pp. 1829 – 1840, Dec. 2000.

3.   P. Balmelli and Q. Huang, "A 25 MS/s 14 b 200 mW $\Sigma\Delta$ modulator in 0.18μm CMOS," *ISSCC Dig. Tech. Papers*, pp. 74 – 514, Feb. 2004.

4.   R. Gaggl, M. Inversi, and A. Wiesbauer, "A power optimized 14-bit SC $\Delta\Sigma$ modulator for ADSL CO applications," *ISSCC Dig. Tech. Papers*, pp. 82 – 514, Feb. 2004.

5.   J. Silva, U. Moon, J. Steensgaard, and G.C. Temes, "Wideband low-distortion delta-sigma ADC topology," *Electron. Lett.*, vol. 37, pp. 737 – 738, 2001.

6.   A. Petragalia and S. K. Mitra, "High speed A/D conversion incorporating a QMF bank," *IEEE Trans. Instrum. Meas.*, vol. 41, pp. 427 – 431, Jun. 1992.

7.   I. Galton and H. T. Jensen, "Oversampling parallel delta-sigma modulation A/D conversion," *IEEE Trans. Circuits Syst. II*, vol. 43, pp. 801 – 810, Dec.1996.

8.   R. Khoini-Poorfard, L.B. Lim, and D.A. Johns, "Time-interleaved oversampling A/D converters: theory and practice," *IEEE Trans. Circuits Syst. II*, vol. 44, no. 8, pp. 634 – 645, Aug. 1997.

9.   M. Kozak and I. Kale, "Novel topologies for time-interleaved delta-sigma modulators," *IEEE Trans. Circuits Syst. II*, vol. 47, no. 7, pp. 639 – 654, Jul. 2000.

10. S. K. Mitra, Digital Signal Processing: A Computer-Based Approach. McGraw-Hill, 2001.

11. R. Schreier, and G. C. Temes, Understanding Delta-Sigma Data Converters, Wiley-IEEE Press, 2004.

12. J. A. Cherry and W. M. Snelgrove, "Excess Loop Delay in Continuous-Time Delta-Sigma Modulators," *IEEE Trans. Circuit and Systems-II: Analog and Digital Signal Processing*, vol. 46, pp. 376-389, Apr. 1999.

13. A. M. Thurston, T. H. Pearce and M. J. Hawksford, "Bandpass Implementation of the Sigma-Delta A-D Conversion Technique," *Int. Conf. on Analogue to Digital and Digital to Analogue Conversion*, pp. 81-86, Sept. 1991.

14. F. M. Gardner, "A Transformation for Digital Simulation of Analog Filters," *IEEE Trans. Communications*, vol. COM-34, pp. 676-680, July 1986.

15. A. Yahia, P. Benabes and R. Kielbasa, "A New Technique for Compensating the Influence of the Feedback DAC Delay in Continuous-Time Band-Pass Delta-Sigma Converters," in *Proc. IEEE Inst. and Meas. Tech. Conf.*, vol. 2, pp. 716-719, 2001.

16. P. Benabes, M. Keramat and R. Kielbasa, "A Methodology for designing Continuous-time Sigma-Delta Modulators," in *Proc. European Design and Test Conf.*, pp. 46-50, 1997.

17. M. Ortmanns, F. Gerfers and Y. Manoli, "Clock Jitter Insensitive Continuous-Time $\Delta\Sigma$ Modulators," in *IEEE Int. Conf. On Electronics, Circuits and Systems*, vol. 2, pp. 1049-1052, 2001.

18. H. Tao, L. Toth and J.M. Khoury, "Analysis of Timing Jitter in Bandpass Sigma-Delta Modulators," *IEEE Trans. Circuits and Systems II*, vol. 46, pp. 991-1001, Aug. 1999.

19. O. Oliaei and H. Aboushady, "Jitter Effects in Continuous-Time $\Delta\Sigma$ Modulators with Delayed Return-to-Zero Feedback," in *IEEE Int. Conf. On Electronics, Circuits and Systems*, vol. 1, pp. 351-354, Sept. 1998.
20. J.A. Cherry and W.M. Snelgrove, "Clock Jitter and Quantizer Metastability in Continuous-Time Delta-Sigma Modulators," *IEEE Trans. Circuits and Systems II*, vol. 46, pp. 661-676, June 1999.
21. A. Tabatabaei, et al., "A Dual Channel $\Delta\Sigma$ ADC with 40MHz Aggregate Signal Bandwidth," *Proc. IEEE ISSCC Dig. Tech. Papers*, pp. 66-67, Feb. 2003.
22. L. J. Breems, "A Cascaded Continuous-Time $\Delta\Sigma$ Modulator with 67dB Dynamic Range in 10MHz Bandwidth," *Proc. IEEE ISSCC Dig. Tech. Papers*, pp. 72-73, Feb. 2004.
23. A. Di Giandomenico, et al., "A 15 MHz Bandwidth Sigma-Delta ADC with 11 Bits of Resolution in 0.13um CMOS," *Conf. on European Solid-State Circuits*, pp. 233-236, Sept. 2003.

International Journal of High Speed Electronics and Systems
Vol. 15, No. 2 (2005) 319–351
© World Scientific Publishing Company

Designing *LC* VCOs Using Capacitive Degeneration Techniques*

Byunghoo Jung and Ramesh Harjani

*Department of Electrical and Computer Engineering,
University of Minnesota, 200 Union Street S.E.
Minneapolis, Minnesota 55455, USA*
†*harjani@ece.umn.edu*

In this paper, we present a detailed analysis of VCOs using a capacitively degenerated negative resistance cell. The negative resistance cell using capacitive degeneration has a higher maximum attainable oscillation frequency and a smaller equivalent shunt capacitance when compared to the widely used cross-coupled negative-g_m cell. These properties are of particular interest for the design of high-frequency and/or wide tuning range VCOs. The negative resistance provided by a traditional capacitively degenerated negative resistance cell is lower than that provided by a cross-coupled negative-g_m cell. We present an active capacitive degeneration topology that overcomes this limitation. To validate this circuit topology we use two test vehicles. The first test vehicle is a 5.3 GHz VCO designed in a 0.25 μm CMOS technology and the second test vehicle is a 20 GHz VCO designed in a 0.25 μm BiCMOS technology. Measurement and simulation results from both test vehicles effectively demonstrate the efficacy of the capacitive degeneration technique.

Keywords: Voltage-controlled oscillators; analog integrated circuits; BiCMOS integrated circuits; capacitive degeneration; high-frequency *LC* oscillators; negative resistance cell.

1. Introduction

Integrated *LC* voltage-controlled oscillators (VCOs) are critical building blocks in high-performance communication systems. The ever-increasing demand for bandwidth places very stringent frequency, power and noise requirements on such systems. Significant research effort has been put into improving the noise and power performance of integrated *LC* VCOs. The majority of this effort is targeted at optimizing the *LC* tank and the negative resistance cell design [1]- [11]. And, bulk of this work is based on the topology using cross-coupled negative resistance cell. The integrated cross-coupled negative resistance cell has been used widely for its simplicity and easy differential implementation. A transistor that has capacitive

---

*Portions of this manuscript and material referred in this manuscript have appeared in [18], [19], and [20].

degeneration can also produce negative resistance. The basic discrete frequency dependent negative resistance cell based on capacitive degeneration has been used previously in [13]- [16], and recently, examples of integrated differential implementations have been presented in [17]- [19] without extensive description of the operating principles and/or comparison with the cross-coupled negative-g_m cell. This paper analyzes the performance and limitations of the different negative resistance cells including the impact of different process technologies.

One of the interesting focus of investigation is the high-frequency behavior of the different negative resistance cells. Even for the widely used cross-coupled negative-g_m cell, there has been very little work on its behavior at frequencies close to the unity current gain frequency, f_T, of the process. Only recently has the maximum attainable oscillation frequency of this topology been presented [12]. The analysis in this paper covers different aspects of the underlying physical mechanisms that limit the high-frequency performance of LC VCOs to provide in-depth design insights for the circuit designers. The analysis considers not only the negative resistance cells but also the design techniques for low-power and low-noise implementations including buffers and driven loads. The analysis shows that, with proper design, the negative resistance cell using capacitive degeneration can out-perform the widely used cross-coupled negative-g_m cell especially at high-frequencies. Another focus of investigation is the equivalent capacitance of the negative resistance cells. This study shows the wide tuning capability of the VCO using a capacitively degenerated negative resistance cell that accrues from its low equivalent capacitance. A design example of a wide tuning range 5.3 GHz VCO in 0.25 µm CMOS technology is provided.

These investigations motivated us to develop a new negative resistance cell topology that can further improve the performance in comparison to previous topologies. The proposed cell is based on the capacitively degenerated topology, but unlike the designs in [17], [18], and [21], it uses two cross-coupled active devices that serve as the degenerating capacitance and also provide additional transconductance. To show the effectiveness of the presented negative resistance cell using an active capacitive degeneration, Section 3.2 presents the design and experimental results for a 20 GHz VCO in IBM SiGe 0.25 µm BiCMOS process, and provides a comparison of its performance with previously reported 20 GHz LC oscillators. This paper also discusses some practical design considerations including the issue of common-mode oscillation and techniques to prevent it, and methods to utilize the small equivalent capacitance of the negative resistance cell to either improve phase noise or increase output signal swing.

2. LC Tank-based VCO Design

Fig. 1 shows a simplified circuit model for a parallel LC oscillator in steady state, where the resistance R_P represents the tank loss, R_{Eq} is the effective negative resistance generated by the active devices, and C_{Eq} is the effective shunt capacitance

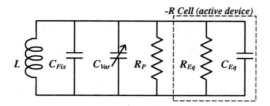

Fig. 1. Parallel LC oscillator model.

contributed by the active devices in the negative resistance cell. For sustained oscillation, the magnitude of the effective negative resistance $|R_{Eq}|$ has to be smaller than R_p. Additionally, for high-frequency operation where C_{Var} is small, the effective capacitance C_{Eq} should be small compared to C_{Var} so as to not limit the maximum attainable oscillation frequency and tuning range.

The traditional cross-coupled cell which is also known as the negative-g_m cell is used widely due to its simplicity and its ability for differential signaling. First of all, we investigate the limitations of the traditional cross-coupled cell in both BiCMOS and CMOS form. We then analyze the maximum attainable frequency of oscillation and show the importance of the base/gate resistance and the equivalent parasitic capacitance (C_{Eq}) of a negative resistance cell. All simulations comparing BiCMOS and CMOS use either the $0.25\,\mu m$ IBM SiGe BiCMOS process or the $0.18\,\mu m$ TSMC CMOS process. Though comparisons between process technologies is never exact, these two processes are comparable with similar f_Ts. Next, we discuss the negative resistance cell using capacitive degeneration. The analysis uses a CMOS example and focuses on the resulting benefits from the small equivalent capacitance (C_{Eq}) of the cell. Lastly, we present a negative resistance cell using an active capacitive degeneration that further improves the performance of the negative resistance cell using a simple capacitive degeneration. The design issues related to this new architecture are discussed using a BiCMOS implementation example. The issue of common-mode oscillation is discussed at the end of this section.

For the rest of this section, we use small-signal models for the behavioral analysis of the different negative resistance cells. Even though small-signal analysis is insufficient to completely predict the large-signal behavior of an oscillator, it still provides designers with good insight and compares fairly well with the simulation results, suggesting the appropriateness of this approach.

2.1. Maximum Attainable Oscillation Frequency for the Cross-Coupled Negative Resistance Cell

In this section we evaluate the maximum frequency of oscillation for a given power budget for BJT and MOS implementations of the cross-coupled structure. Several factors (negative resistance cell design, buffer stage bandwidth, frequency tuning range and the parasitic capacitances of the constituent components) affect the maximum oscillation frequency. Due to its higher transconductance for a given current,

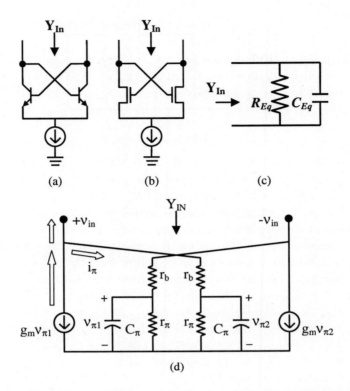

Fig. 2. Cross-coupled negative resistance cell: (a) BJT, (b) MOS , (c) equivalent shunt model, (d) and differential small-signal model for a BJT implementation.

the BJT implementation provides a better R_{Eq}. However, as will be shown later, the finite base resistance severely limits the maximum attainable oscillation frequency.

2.1.1. *Negative resistance cell*

We first consider the mechanisms that degrade high-frequency performance in a cross-coupled negative resistance cell. Veenstra [12] derived expressions for the maximum attainable oscillation frequency for a BJT cross-coupled cell using small-signal analysis. We use a simpler method that provides better physical insight into the degradation mechanisms. Fig. 2 shows simplified circuit schematics for the (a) BJT-based and (b) CMOS-based cross-coupled cells, (c) the shunt equivalent $R_{Eq} \| C_{Eq}$ model, and (d) the differential small-signal model for the BJT version. At low frequencies, R_{Eq} is approximately $-2/g_m$ where g_m is the device transconductance. Typically BJTs have a higher g_m than CMOS transistors for a given bias current and are a better choice for low power design. Additionally, the low flicker noise corner makes BJT devices better candidates for low-power low-noise oscillator designs. But, with increasing frequency, the magnitude of $|R_{Eq}|$ becomes large and eventually flips over and R_{Eq} becomes positive. This is primarily due to the non-negligible base resistance, r_b, at high frequencies.

At high frequencies, the impedance offered by the base-emitter capacitance C_π becomes comparable to r_b as a result a larger portion of the base voltage appears across r_b rather than C_π. Additionally, there is a phase difference between the voltage across C_π and the base voltage due to the resistor-capacitor voltage division. The voltage across C_π can be calculated using the small-signal model shown in Fig. 2 (d). Using the voltage division ratio (v_π/v_{in}), we can define an effective transconductance:

$$G_{m,eff} \equiv \frac{v_\pi}{v_{in}} \equiv A\,g_m \tag{1}$$

In Fig. 2 (d) the difference between the transconductor current and i_π is returned to the signal source, thereby generating a negative resistance. A relationship between the current i_π and the transconductance associated with this current (g_π) is given by $g_\pi \equiv i_\pi/v_{in}$. Only the real parts of $G_{m,eff}$ and g_π contribute to the equivalent negative resistance. Therefore, R_{Eq} can be approximated as:

$$R_{Eq} \approx \frac{-2}{\text{Re}\,[G_{m,eff}] - \text{Re}\,[g_\pi]} \tag{2}$$

Equation (2) suggests that R_{Eq} degrades rapidly as $\text{Re}[g_\pi]$ approaches $\text{Re}[G_{m,eff}]$, and has a distinct corner frequency which will be revisited later. An expression for the negative to positive transition frequency of R_{Eq} (ω_{tran}) can be calculated by considering the condition $\text{Re}[G_{m,eff}] - \text{Re}[g_\pi] \leq 0$, which results in:

$$\omega_{tran} = \sqrt{\frac{\left(1 + \frac{r_b}{r_\pi}\right)\left(g_m - \frac{1}{r_\pi}\right)}{r_b C_\pi^2}} \tag{3}$$

Assuming $r_b/r_\pi \ll 1$, $r_\pi = \beta/g_m$ and $\beta \gg 1$ where β is the base to collector current gain, then equation (3) can be simplified to:

$$\omega_{tran} \approx \sqrt{\frac{\omega_T}{r_b C_\pi}} \tag{4}$$

where $\omega_T = g_m/C_\pi$. For the MOS version of the cross-coupled structure, r_b and C_π are replaced with r_g and C_{gs}. Equation (4) shows that the transition frequency is a strong function of the bias current, device size and base resistance. In the case of the MOS transistor, a multiple finger layout can be used to drastically minimize gate resistance [22]. The relatively high base resistance of the BJT in comparison to the gate resistance of the CMOS device can result in a lower transition frequency, despite its higher ω_T. Fig. 3 (a) shows the simulated R_{Eq} and C_{Eq} for the BJT and MOS cross-coupled cells using the 0.25 μm BiCMOS and 0.18 μm CMOS technologies respectively at a bias current of 2 mA. We see that the BJT-based design shows a smaller $|R_{Eq}|$ than its CMOS counterpart at low frequencies due to its higher transconductance value. However, it has a much lower transition frequency (~ 28 GHz) as compared to the CMOS case (~ 53 GHz). Fig. 3 (b) shows plots for

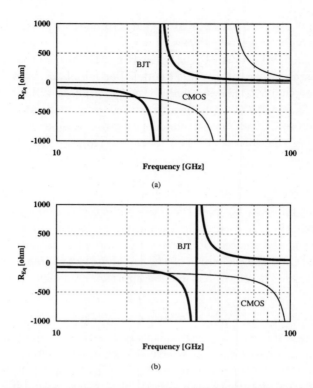

Fig. 3. Negative resistance for the cross-coupled cell (a) simulation and (b) model.

$|R_{Eq}|$ for the BJT and MOS cross-coupled structures using equation (2). We see that equation (2) predicts a transition frequency that is higher than seen in the simulation results. This is primarily because we neglected the emitter resistance, base-collector capacitance, and emitter-collector capacitance and resistance in our analysis. But more importantly, it correctly predicts the rapid degradation of R_{Eq} after the corner frequency. Increasing the bias current increases ω_T, the transition frequency. This implies that a low base (gate) resistance design is important not only from a noise perspective but also from a low-power perspective. This analysis suggests that careful layout is imperative to increase the transition frequency, especially for the CMOS case where the gate resistance can be reduced substantially by appropriate layout techniques without any serious impact on ω_T and/or C_{gs}.

2.1.2. *Buffer stage design*

Another important factor that impacts the maximum attainable frequency of oscillation is the buffer stage. Buffer stages are usually implemented as emitter followers because of their high input impedance and wide bandwidth. Fig. 4 shows the simplified circuit schematic for the VCO core coupled with a buffer. The input admittance,

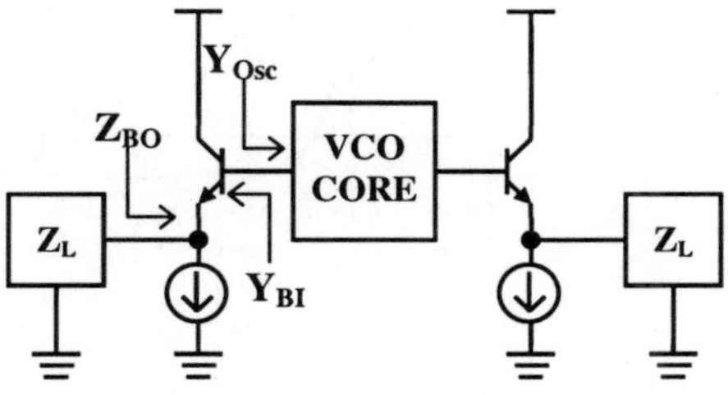

Fig. 4. Emitter follower buffer stage schematic.

Y_{BI}, of the buffer stage can be written as:

$$Y_{BI} = \frac{1}{Z_L \left[1 + \frac{r_\pi}{1 + j\omega C_\pi r_\pi} \left(g_m + \frac{1}{Z_L}\right)\right]} \tag{5}$$

When the load impedance is resistive, $Z_L = R_S$, the effective input shunt resistance and capacitance looking into the buffer are given by equations (6) and (7).

$$R_{Eq,BI} = R_S \frac{\left[1 + \beta\left(1 + \frac{1}{g_m R_S}\right)\right]^2 + \left(\beta \frac{\omega}{\omega_T}\right)^2}{1 + \beta\left(1 + \frac{1}{g_m R_S}\right) + \left(\beta \frac{\omega}{\omega_T}\right)^2} \approx R_S \left[1 + \left(\frac{\omega_T}{\omega}\right)^2 \left(1 + \frac{1}{g_m R_S}\right)^2\right] \tag{6}$$

$$C_{Eq,BI} = C_\pi \frac{r_\pi^2 \left(\frac{g_m}{R_S} + \frac{1}{R_S^2}\right)}{\left[1 + r_\pi\left(g_m + \frac{1}{R_S}\right)\right]^2 + (\omega C_\pi r_\pi)^2} \approx \frac{1}{\omega_T R_S} \frac{\left(1 + \frac{1}{g_m R_S}\right)}{\left(1 + \frac{1}{g_m R_S}\right)^2 + \left(\frac{\omega}{\omega_T}\right)^2} \tag{7}$$

If we assume that $\beta \gg 1$ and $\omega \sim \omega_T$, then we can make the approximations shown in equations (6) and (7). It is essential that $R_{Eq,BI}$ is large enough so as to not reduce the loaded Q of the tank, while $C_{Eq,BI}$ is small enough so as to not limit the frequency tuning range. In many integrated transceiver designs the buffer drives an internal capacitive load, C_L. In this case the effective shunt resistance and capacitance looking into the buffer input are as follows:

$$R_{Eq,BI} = -\frac{(1 + g_m r_\pi)^2 + \omega^2 r_\pi^2 (C_\pi + C_L)^2}{\omega^2 r_\pi C_L (g_m r_\pi C_\pi - C_L)} \tag{8}$$

$$C_{Eq,BI} = \frac{(1 + g_m r_\pi) C_L + \omega^2 r_\pi^2 C_\pi C_L (C_\pi + C_L)}{(1 + g_m r_\pi)^2 + \omega^2 r_\pi^2 (C_\pi + C_L)^2} \tag{9}$$

Equation (8) predicts a negative input resistance at the buffer input when $g_m r_\pi > C_L/C_\pi$. This condition is easily met as $g_m r_\pi (= \beta)$ can easily be greater than ten. This additional negative resistance provided by the buffer can be utilized by the VCO core when driving a capacitive load. Fig. 5 shows the small-signal model for the VCO core and buffer load. The combined negative resistance is equal to

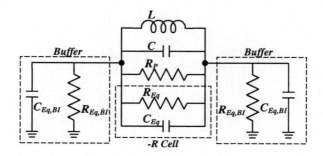

Fig. 5. Parallel LC oscillator model including buffer stage impedance.

$R_{Eq}\|2R_{Eq,BI}$. The negative resistance provided by the buffer reduces the g_m requirement of the cross-coupled cell, and hence can be used to reduce power consumption and noise. Equation (9) predicts that the effective shunt capacitance is smaller than C_π or C_L which allows for a high-frequency design with wider tuning range. This basic negative resistance seen in the buffer due to capacitive emitter degeneration will be used to build a new negative resistance cell for the VCO in the following sections.

The impedance looking at the output of the buffer, Z_{BO}, determines the bandwidth of the buffer. In a common base configuration, the output impedance is equal to $\sim 1/g_{m,buffer}$ at low frequencies. However, when driven by a high impedance oscillator, Y_{OSC} affects Z_{BO} as shown in Fig. 4. The output impedance for this configuration is given by:

$$Z_{BO} = \frac{Z_\pi + Z_{OSC}}{g_{m,buffer}Z_\pi + 1} \tag{10}$$

where $Z_\pi = (1/sC_\pi \parallel r_\pi)$ and $Z_{OSC} = 1/Y_{OSC}$. In Figure 6 we show circuit and bias details for MOS and bipolar oscillator cores. The oscillator admittance, Y_{OSC} is a strong function of Z_{biasL} and Z_{biasC} as well as Y_{BI}. Nevertheless, because the real part of Z_{BO} tends to increase with an increasing base impedance, we can derive an optimistic pole frequency produced at the emitter of the buffer stage given by:

$$\omega_{Pole} = \frac{g_{m,buffer}}{C_L} \tag{11}$$

This pole frequency has to be higher than the oscillation frequency to prevent output signal attenuation. More importantly, the signal suffers fairly large phase shift around the pole frequency, and relatively small process variations can introduce large phase shift between two output signals when the pole frequency is close to the oscillation frequency. In Fig. 7 we plot the pole frequency, ω_{Pole} as a function of the load capacitance C_L at the output of the buffer stage for bipolar and MOS implementations at a 1mA bias current. The MOS design has a lower pole frequency even with a relatively small load capacitance due to its lower g_m. This implies that even though the MOS cross-coupled cell can provide a negative resistance up to very

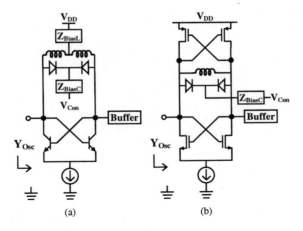

Fig. 6. Oscillator output admittance examples for (a) BJT, (b) MOS VCOs.

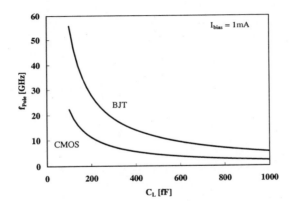

Fig. 7. Pole frequency vs. load capacitance of the buffer stage.

high frequencies, the smaller bandwidth of the buffer stage can limit the practical maximum attainable oscillation frequency.

2.1.3. *Parasitic capacitance and tuning range*

Another important factor that determines the maximum attainable oscillation frequency for a given technology and power budget are the parasitic capacitances of the various components. The maximum oscillation frequency of the circuit in Fig. 1 is $1/\sqrt{L(C_{Fix} + C_{Eq})}$. When a desired tuning range, $\Delta\omega$, is required, the maximum attainable center frequency can be described as:

$$\omega_{Max} = \frac{1}{\left(1 + \frac{Tune}{200}\right)\sqrt{L_{Min}\left(C_{Fix} + C_{Eq}\right)}} \tag{12}$$

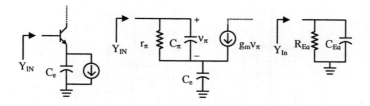

Fig. 8. Single ended negative resistance cell using capacitive emitter degeneration.

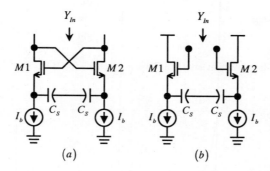

Fig. 9. Circuit schematics for the differential negative resistance cells using capacitive degeneration in CMOS implementation.

where $Tune$ is the tuning range which is expressed as a percentage of the center frequency, L_{Min} is the smallest feasible inductance without suffering severe process variations, C_{Fix} is the fixed sum of parasitic capacitances contributed by the varactor, the inductor, and the buffer, and C_{Eq} is the effective shunt capacitance of the negative resistance cell. The effective capacitance of the cross-coupled cell is $\sim C_\pi/2$. At high frequencies, this along with the effective capacitance from the buffer stage becomes a significant portion of the overall capacitance and limits the maximum attainable oscillation frequency and tuning range.

In this subsection, we showed the mechanisms that degrade the R_{Eq} generated by the cross-coupled cell and showed how the effective capacitance from the buffer and negative resistance cell limits the maximum attainable frequency and tuning range. These limitations have motivated a new negative resistance cell, described in the next subsection, that has a higher negative to positive transition frequency and lower effective capacitance thereby enabling low-power low-noise high-frequency design. Even though the new design is technology independent, the benefits are maximized in a BiCMOS implementation, and hence the following sections will focus on a BiCMOS design.

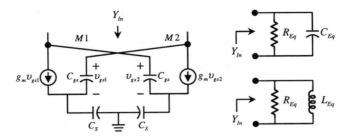

Fig. 10. The small signal model and the equivalent shunt models for the capacitively degenerated negative resistance cell.

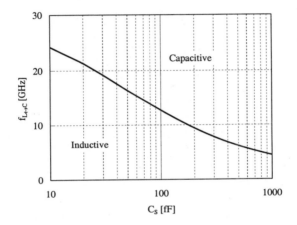

Fig. 11. The transition frequency vs. C_S.

2.2. *Negative Resistance Cell Using Capacitive Degeneration*

As seen in the previous section, a capacitive emitter/source degenerated transistor can generate negative resistance. Resonator-based VCOs using this technique have been presented in [13]- [21], and a similar topology has been used in a relaxation oscillator [23]. Fig. 8 shows a single ended negative resistance cell using capacitive emitter degeneration. The R_{Eq} and C_{Eq} can be described using the formulas in equations (8) and (9). When a voltage is applied to the base, a part of it appears across C_π, which in turn generates the transconductor current. Portion of this transconductor current is returned to the signal source via the capacitor-capacitor divider. A portion of this returning current has a 180^0 phase shift from the voltage that is applied, which signifies a negative resistance. This negative resistance cell has a very small effective capacitance. First of all, C_π and C_e are in series. Second, the g_m-cell affects the charge buildup across each capacitor and increases the effective C_π while decreasing the effective C_e, thus further reducing the effective series capacitance. This smaller effective parasitic capacitance compared to that

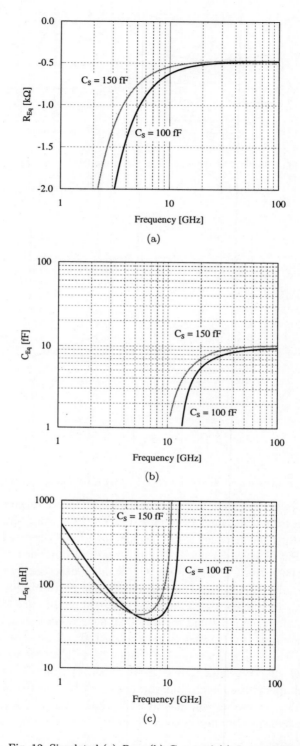

Fig. 12. Simulated (a) R_{Eq}, (b) C_{Eq}, and (c) L_{Eq} results.

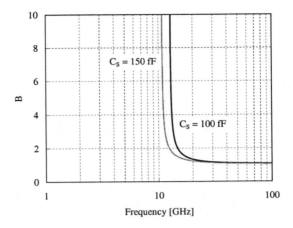

Fig. 13. Benefit factor vs. frequency.

of cross-coupled negative-g_m cell, $\sim C_\pi/2$ and $\sim C_{gs}/2$ in Bipolar and CMOS respectively, enables the design of wide tuning range VCO. Fig. 9 shows the CMOS implementation of differential negative resistance cells using capacitive degeneration. The negative resistance cell shown in Fig. 9(a) has cross-link. For the negative resistance cell without the cross-link, as shown in Fig. 9(b), the LC tank can be connected between the gates, and the power supply can be connected to the drains of the transistors. This connection can improve the power supply pulling and supply noise rejection. But, the cell without cross-link is susceptible to common-mode oscillations, and care has to be taken to prevent this from occurring. These mechanisms will be discussed at the following sections. In this section, we focus on the cell with cross-link, but the cell without cross-link shows similar behavior. Because the tank is connected in shunt across the negative resistance cell, an equivalent shunt resistance and shunt reactance model for negative resistance cell is convenient for analysis. Fig. 10 shows the small signal model and the equivalent shunt model of the negative resistance cell using capacitive degeneration. For the negative resistance cell with capacitive source degeneration, the imaginary part of the equivalent shunt impedance can become inductive instead of capacitive. Fig. 10 shows simplified models for both the capacitive and the inductive cases. Equation (13) shows the calculated input admittance, and equations (14), (15) and (16) show the extracted equivalent resistance (R_{Eq}), capacitance (C_{Eq}), and inductance (L_{Eq}) respectively where ω_T is g_m/C_{gs}.

$$Y_{In} = \frac{1}{2\left[g_m^2 + \omega^2\left(C_{gs} + C_s\right)^2\right]} \times \left\{ \begin{array}{l} -\omega^2 C_s \left(2C_{gs} + C_s\right) g_m \\ +j\omega C_s \left[-g_m^2 + \omega^2 C_{gs}\left(C_{gs} + C_s\right)\right] \end{array} \right\} \quad (13)$$

$$R_{Eq} = \frac{-2\left[1 + \left(\frac{\omega}{\omega_T}\right)^2 \left(1 + \frac{C_s}{C_{gs}}\right)^2\right]}{g_m \left(\frac{\omega}{\omega_T}\right)^2 \frac{C_s}{C_{gs}} \left(2 + \frac{C_s}{C_{gs}}\right)} \qquad (14)$$

$$C_{Eq} = \frac{C_S \left[-1 + \left(\frac{\omega}{\omega_T}\right)^2 \left(1 + \frac{C_s}{C_{gs}}\right)\right]}{2\left[1 + \left(\frac{\omega}{\omega_T}\right)^2 \left(1 + \frac{C_s}{C_{gs}}\right)^2\right]} \qquad (15)$$

$$L_{Eq} = \frac{2\left[1 + \left(\frac{\omega}{\omega_T}\right)^2 \left(1 + \frac{C_s}{C_{gs}}\right)^2\right]}{\omega^2 C_s \left[1 - \left(\frac{\omega}{\omega_T}\right)^2 \left(1 + \frac{C_s}{C_{gs}}\right)\right]} \qquad (16)$$

The imaginary part provides inductive impedance up to a certain frequency and then changes into capacitive impedance. Equation (17) shows the transition frequency where the impedance changes from being inductive to being capacitive.

$$\omega_{L \leftrightarrow C} = \omega_T \sqrt{\frac{C_{gs}}{C_{gs} + C_s}} \qquad (17)$$

Fig. 11 shows the transition frequency as a function of the source degeneration capacitor C_s. As C_s increases, the transition frequency decreases. For typical value of C_s and C_{gs}, $\omega_{L \leftrightarrow C}$ is about 0.4-0.5 ω_T. The simulated values for R_{Eq}, C_{Eq}, and L_{Eq} using small signal analytical calculations are shown in Figs. 12(a), 12(b) and 12(c) respectively. The negative resistance for the proposed structure is smaller than that for a typical cross-coupled negative-g_m structure, but the difference reduces with increasing C_s. Below the inductive to capacitive transition frequency the effective shunt inductance is fairly large and has minimal effect on the inductance value in the tank. The tradeoff between the reduced negative resistance and effective parasitic capacitance has to be optimized. To compensate for the smaller negative resistance of the proposed structure, the g_m has to be increased. For a given power budget, this can be done by increasing the transistor size in CMOS. This increases the effective parasitic capacitance, but as long as the increased effective parasitic capacitance of the capacitively degenerated negative resistance cell is less than the effective parasitic capacitance of the traditional cross-coupled negative-g_m cell, the proposed design is better. Because the g_m is proportional to the square root of the device size, and the R_{Eq} is inversely proportional to g_m for typical structure, we can define a benefit factor, B, as done in equation (18).

$$B = \frac{1}{R_{ratio}\sqrt{C_{ratio}}} \qquad (18)$$

where

$$C_{ratio} = \frac{C_{EQ}(New)}{C_{EQ}(Typical)} \qquad (19)$$

and

$$R_{ratio} = \frac{R_{EQ}\,(New)}{R_{EQ}\,(Typical)}. \tag{20}$$

Fig. 13 shows the benefit factor as a function of frequency. The benefit factor is small near ω_T but increases sharply with decreasing frequency. This implies that for a given $-R_{Eq}$ requirement and power budget, the proposed structure has a smaller parasitic capacitance; such that it is capable of a larger tuning range or a larger output swing. However, care should be taken to ensure that too large a transistor size is not selected to limit the actual benefit.

2.3. *Active Capacitive Degeneration Based Negative Resistance*

Another unique property of the negative resistance cell using capacitive degeneration is a high negative to positive transition frequency of R_{Eq} as presented in [17]. The presentation in [17] does not explain the physical mechanism for this high transition frequency. The small effective capacitance has a higher impedance than C_π resulting in a smaller effective r_g. This causes the voltage drop across r_g and the phase shift from the base voltage to the voltage across C_π to be smaller than in the cross-coupled cell. So, it retards the R_{Eq} degradation mechanism, and increases the R_{Eq} transition frequency. However, emitter/source degeneration decreases the effective transconductance. Because the R_{Eq} transition frequency is also a function of g_m, the reduced effective transconductance limits the improvement in the R_{Eq} transition frequency and also increases the $|R_{Eq}|$ value. For a given power budget, increasing the device size can compensate the increased $|R_{Eq}|$ in CMOS technology as shown in previous section. But in BJT where the g_m is insensitive to device size to the first order, the increased $|R_{Eq}|$ requires more current to increase g_m, or a higher Q tank design, and hence higher Q inductor design which is not trivial at high frequencies. In this section, we present a further improvement to the capacitively emitter degenerated negative resistance cell in BiCMOS implementation that mitigates the reduction in g_m issue mentioned here. Fig. 14 shows the evolution of the proposed negative resistance cell. Two capacitively emitter degenerated negative resistance cells are interconnected by using a cross-coupled NMOS pair. The gate and drain of each NMOS device is connected to the emitters of each negative resistance cell respectively. The C_{gs} of the NMOS device replaces the emitter degenerating capacitor, and two current sources are combined into one as shown in Fig. 14. The cross-coupled NMOS pair works both as emitter degenerating capacitors and also provides additional transconductance to improve the negative resistance value. Fig. 15 shows the equivalent small signal circuit model for the proposed negative resistance cell. The input admittance is calculated to be:

$$Y_{IN} = \frac{\frac{1}{2}\left[g_{m2}\left(\frac{r_g}{Z_g}-1\right)+\frac{1}{Z_g}\right]}{[1+g_{m1}Z'_\pi]+Z_\pi\left[g_{m2}\left(\frac{r_g}{Z_g}-1\right)+\frac{1}{Z_g}\right]} \tag{21}$$

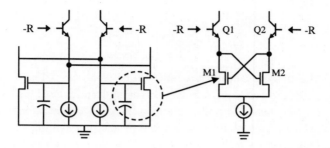

Fig. 14. Differential negative resistance cell using capacitive degeneration and cross-coupled MOS transistors

where $Z'_\pi = (r_\pi \parallel 1/sC_\pi)$, $Z_\pi = r_b + Z'_\pi$, $Z_g = r_g + 1/sC_{gs}$, and g_{m1} and g_{m2} are transconductances for the BJT and the NMOS transistor respectively. If we ignore r_b and r_g the equivalent shunt resistance is given by:

$$R_{Eq} = \frac{-2\left[\left(\frac{1}{r_\pi} + \Delta g_m\right)^2 + (\omega C_T)^2\right]}{\left(\frac{1}{r_\pi} + \Delta g_m\right)\left(\frac{g_{m2}}{r_\pi} + \omega^2 C_{gs} C_\pi\right) - \omega^2 C_T \left(\frac{C_{gs}}{r_\pi} - C_\pi g_{m2}\right)} \qquad (22)$$

where $\Delta g_m = g_{m1} - g_{m2}$ and $C_T = C_\pi + C_{gs}$. For a fixed bias current, the transconductance of the BJT is to first order unaffected by the device geometry, but the transconductance of MOS device increases as the square root of the device size. To find the optimum g_{m1}/g_{m2} ratio, we let $g_{m1} = \delta g_{m2}$, and $g_{m2} = K\sqrt{C_{gs}}$, where K is a constant. Using this relationship and assuming that $g_{m1} r_\pi \gg 1$ then equation (22) can be simplified to equation (23).

$$R_{Eq} = \frac{-\frac{2}{\delta}\left[(1-\delta)^2 + \left(\frac{\omega}{\omega_T}\right)^2\left(1 + \frac{\Gamma^2}{C_\pi}\right)\right]}{(1-\delta) g_{m1}\left(\frac{1}{\beta} + \frac{\omega^2\Gamma}{\omega_T K}\right) - \omega^2\left(\frac{\Gamma}{\beta K} - \frac{1}{\omega_T}\right)(C_\pi + \Gamma^2)} \qquad (23)$$

where $\omega_T = g_{m1}/C_\pi$, and $\Gamma = \delta g_{m1}/K$. In Fig. 16 we plot the expression in equation (23) as a function of δ for different values of g_{m1}. It shows that the R_{Eq} is optimized when δ is near unity, i.e., $g_{m1} \approx g_{m2}$. Additionally, it shows that this ratio is more critical than the magnitude of g_{m1} itself. For example a 15mA/V-15mA/V combination results in a better R_{Eq} than a 30mA/V-15mA/V combination which gives some important design insights for optimizing the design.

Fig. 17 and 18 show the simulated R_{Eq} and C_{Eq} results for the cross-coupled cell, the simple capacitively degenerated cell, and the proposed active capacitively degenerated cell respectively. All designs use 2mA bias current and use the same BJT size. As predicted the simple capacitive degeneration cell shows a higher R_{Eq} transition frequency but a larger $|R_{Eq}|$ in comparison to the cross-coupled cell. The proposed cell also has a higher R_{Eq} transition frequency in comparison to the cross-coupled cell, but it shows a better R_{Eq} in comparison to the simple capacitively

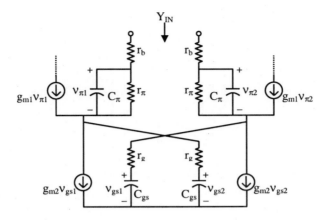

Fig. 15. Equivalent small-signal model for the proposed topology

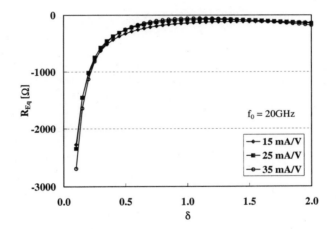

Fig. 16. R_{Eq} vs. δ for different values of g_{m1}.

degenerated cell. This implies that the proposed cell is particularly useful for low-power low-noise design for frequencies above the R_{Eq} transition frequency of the cross-coupled negative-g_m cell. This is because it does not require an increased bias current for a proportional increased g_m which would also increase the noise from the transistor. Fig. 18 shows that the proposed cell has a little higher C_{Eq} in comparison with the simple capacitively degenerated cell between 20GHz and 40GHz, but this value is much smaller than the C_{Eq} of the cross-coupled negative-g_m cell.

The analysis results in Fig. 16 suggest that a BJT-BJT combination would be preferred to a BJT-MOS combination for better g_m matching for the proposed cell. Even though a BJT-BJT combination results in a better R_{Eq}, the relatively high r_b of the BJT results in a low R_{Eq} transition frequency and its benefit reduces to

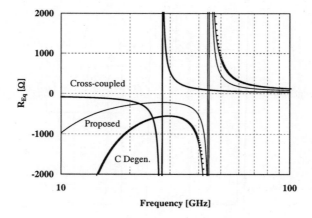

Fig. 17. Simulated R_{Eq} for the cross-coupled, capacitive degeneration and proposed cells.

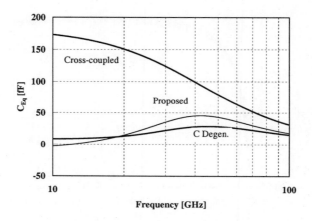

Fig. 18. Simulated C_{Eq} for the cross-coupled, capacitive degeneration and proposed cells.

only the small effective capacitance.

2.4. *Common-Mode And Differential Mode Oscillation*

One distinct difference between the capacitively degenerated cell and the cross-coupled cell is its common-mode behavior. The common-mode input admittance for the ideal cross-coupled cell in Fig. 2 is near zero ($Y_{IN} \approx 0$). Fig. 19 shows the common-mode equivalent circuit schematic for the capacitively degenerated cell. Unlike the cross-coupled cell, each half circuit of the capacitively emitter degenerated cell provides negative resistance. For common-mode signals, the emitter degeneration capacitor connected between two emitters in series (C_S) does not have any effect on the common-mode impedance, only the capacitors connected to

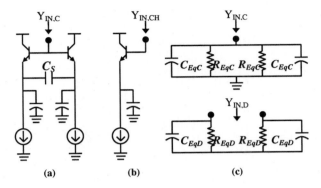

Fig. 19. Common mode circuit equivalent for the capacitively emitter degenerated cell.

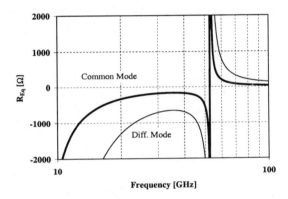

Fig. 20. Common mode negative resistance simulation results.

the ground act as degeneration capacitors. Therefore, when $C_S = 0$, each half of the circuit has the same effective resistance and capacitance as in the differential mode, i.e. $R_{EqD} = R_{EqC}$ and $C_{EqD} = C_{EqC}$. And in this case, the differential and common-mode admittances can be described as:

$$
\begin{aligned}
R_{EqH} &= R_{EqC} = R_{EqD}, \\
C_{EqH} &= C_{EqC} = C_{EqD}, \\
Y_{IN,C} &= \frac{2}{R_{EqH}} + j\omega 2C_{EqH}, \\
Y_{IN,D} &= \frac{1}{2R_{EqH}} + j\omega \frac{C_{EqH}}{2}.
\end{aligned}
\tag{24}
$$

The equations above predict that the common mode $|R_{Eq}|$ is 4 times smaller than the differential mode $|R_{Eq}|$, and common mode C_{Eq} is 4 times larger than the differential mode C_{Eq}. It implies that the capacitively emitter degenerated cell is more prone to oscillate in common-mode when $C_S = 0$.

Fig. 20 shows the simulated common-mode and differential mode R_{Eq} for a capacitively emitter degenerated cell where $C_S = 0$. As expected the common mode shows a much smaller $|R_{Eq}|$. When we use this type of negative resistance cell, the

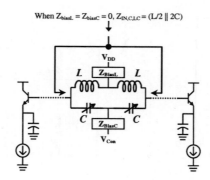

Fig. 21. Common-mode tank schematic with bias network impedance.

tank bias network has to be carefully designed to prevent common-mode oscillation. Fig. 21 shows the LC tank with a center tapped inductor and capacitor including the bias network. If Z_{biasL} and Z_{biasC} are very small, the common mode impedance to ground becomes $(L/2 \parallel 2C)$. And circuit can now oscillate at $\omega = 1/\sqrt{L(C + C_{EqH})}$ in common-mode. In fact, when Z_{biasL} and Z_{biasC} are zero, each half circuit becomes an independent oscillator, and there is no phase relationship constraint between two output signals. If Z_{biasL} and Z_{biasC} are very large, the common-mode impedance of the tank becomes very large, and the circuit cannot oscillate in common-mode. If the bias network impedance is zero, the circuit prefers to oscillate in common-mode. Thus, a bias network with large impedance converts the oscillation mode from common-mode to differential-mode. Fig. 22 shows the oscillation mode conversion from common mode to differential mode for the same circuit by simply altering the bias network impedance. The differential mode signal is shifted down in this plot for graphical convenience only.

In the proposed negative resistance cell with the cross-coupled MOS pair, the common-mode admittance of the MOS pair is near zero. Only the effective input capacitance of the buffer contributes to the common-mode negative resistance. This makes the proposed cell less suspectable to common-mode oscillations as compared to a simple capacitively emitter degenerated cell.

2.5. *Wide Tuning Range*

A wide tuning range is often important because it can accommodate more process and temperature variations. For a high frequency oscillator design, the effective capacitance from the negative resistance cell is one of the key factors that limits the tuning range. The proposed capacitively emitter degenerated cell has a small effective capacitance. Fig. 23 graphically shows the mechanism that results in a small effective capacitance. In the cross-coupled negative-g_m structure, the tank looks into the base of $Q2$, the collector of $Q1$, and the base of buffer transistor $Q3$.

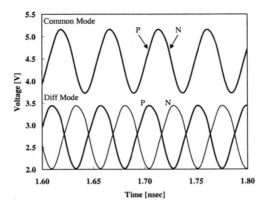

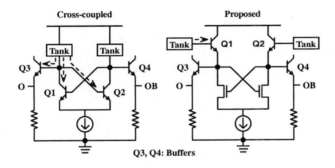

Fig. 22. Common-mode and differential mode oscillation simulations.

Fig. 23. Tank loading characteristics of the cross-coupled cell vs. the proposed cell.

But in the active capacitively degenerated topology, the tank only looks into the base of $Q1$. Furthermore, the equivalent capacitance looking into the base of $Q1$ is much smaller than the C_π of $Q1$ as shown in Section 2.3.

Fig. 24 shows how the effective capacitance of a negative resistance cell affects the oscillator tuning range. For this simulation, a two turn spiral inductor with 740 pH inductance and 17 fF parasitic capacitance, and diode varactors with capacitance range from 28.6 fF to 68.4 fF are used. The tuning range and center frequency go down rapidly with increasing parasitic capacitance of the negative resistance cell. In particular, if we keep the center frequency constant by changing the varactor size, the tuning range reduces even more rapidly as shown in Fig. 24. For example, if the load at the output port in Fig. 23 is 200 fF, the equivalent shunt capacitance looking into the base of $Q3$ is about 44.4 fF. This parasitic loading causes severe frequency and tuning range limitations for the cross-coupled negative-g_m topology as can be seen in Fig. 24. But in the proposed topology, the equivalent shunt capacitance looking into the base of $Q3$ functions as part of the emitter degenerating capacitance. And the effective parasitic capacitance loading the tank is further attenuated

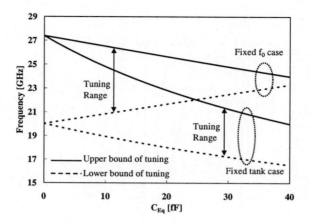

Fig. 24. Simulated tuning range vs. effective capacitance of negative resistance cell.

by $Q1$.

A small effective capacitance also provides more room to increase the inductor size in the tank. In many cases the R_p of LC tank increases with increasing inductance value [24]. So, a large inductance value enables a large output signal power. Depending on the technology and the design constraint, the Q of inductor can also increase with increasing inductance [25]. In this case, both the increased signal power and improved Q contributed by the small effective capacitance can result in lower phase noise as can be seen in the modified Leeson's phase noise formula [26] shown in equation (25).

$$L\left(\Delta f, K_{VCO}\right) = 10 \log\left\{\left(\frac{f_0}{2Q\Delta f}\right)^2\left[\frac{FkT}{2P_0}\left(1+\frac{f_C}{\Delta f}\right)\right] + \frac{1}{2}\left(\frac{K_{VCO}v_m}{2\Delta f}\right)^2\right\} \quad (25)$$

where

$L\left(\Delta f, K_{VCO}\right)$	phase noise in dBc/Hz;
f_0	frequency of oscillation in Hz;
Δf	frequency offset from the carrier in Hz;
F	noise figure of the transistor amplifier;
k	Boltzmann's constant in J/K;
T	temperature in K;
P_0	RF power produced by the oscillator in W;
f_C	flicker noise corner frequency in Hz;
K_{VCO}	gain of the VCO in Hz/V;
v_m	total amplitude of all low frequency noise sources in V/$\sqrt{\text{Hz}}$

The first term within the *log* in equation (25) predicts a decreased phase noise with increasing tank Q and signal power. This discussion emphasizes that the small effective capacitance of the negative resistance cell is not only advantageous for high frequency and wide tuning designs, but results in additional flexibility during

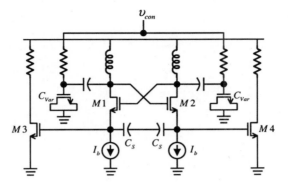

Fig. 25. Circuit schematic of the designed VCO.

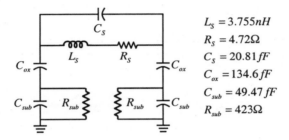

$$L_S = 3.755nH$$
$$R_S = 4.72\Omega$$
$$C_S = 20.81\,fF$$
$$C_{ox} = 134.6\,fF$$
$$C_{sub} = 49.47\,fF$$
$$R_{sub} = 423\Omega$$

Fig. 26. Equivalent inductor model used in the example design.

the LC tank phase noise optimization process.

3. Design Examples

This section presents two design examples of integrated differential VCOs using the capacitive degeneration technique. The 5.3 GHz VCO in the 0.25 μm CMOS technology demonstrates the wide tuning capability of the negative resistance cell using capacitive degeneration. While the 20 GHz VCO in the 0.25 μm SiGe BiCMOS technology illustrates the effectiveness of the negative resistance cell using active capacitive degeneration at high frequencies.

3.1. *5.3GHz VCO in CMOS*

The TSMC 0.25 μm CMOS technology has been used to design one 5.3 GHz VCO. Fig. 25 shows the circuit schematic for the VCO. Bias details are not included. Three and half turn inductor providing 3.715 nH of inductance, 4.715 Ω series resistance (R_S) and 134.6 fF parasitic capacitance is connected to the drain of each transistor. The equivalent inductor model [27] used in this design is shown in Fig. 26.

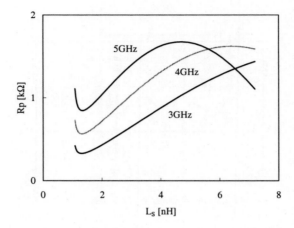

Fig. 27. Equivalent parallel resistance vs. inductance.

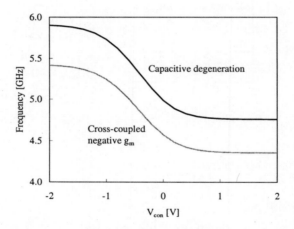

Fig. 28. Tuning range simulation results.

At 5.3 GHz the inductor Q is around 15. Assuming the tank Q is dominated by inductor Q, the equivalent parallel resistance at resonance frequency is roughly $R_S(1+Q^2) = 1.1 K\Omega$. So the negative resistance cell should have R_{EQ} greater than $-1.1\ K\Omega$ and the design target was $-900\ \Omega$. With 3.715 nH inductance, the required capacitance for 5.3 GHz oscillation is 121 fF. A MOS capacitor has been used as a varactor. A MIM capacitor is inserted between drain and MOScap to apply control voltage. Combined with the MIM cap, the fixed portion of the variable capacitance is 65 fF and the inductor has an equivalent parasitic capacitance of 25 fF at 5.3 GHz. Connecting the buffer to the source of transistor as shown in Fig. 25 removes the effect of buffer parasitic capacitance that is not possible in the typical cross-coupled

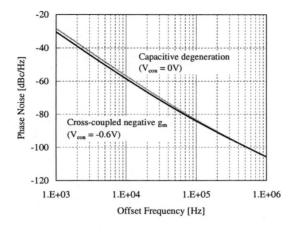

Fig. 29. Phase noise simulation results, $f_0 = 5\,\text{GHz}$.

negative-g_m structure. Because the source follower buffer offers a capacitive impedance, the complete replacement of C_s with the buffer is possible. The negative resistance cell has been designed to have negligible effective parasitic capacitance. So by fully utilizing the tunable portion of the capacitance, 60 fF, an oscillator centered at 5.3 GHz can be designed with a with a significant tuning range. The effective parasitic capacitance for a cross-coupled negative-g_m cell, exceeding several tens of femto farads, combined with the buffer parasitic capacitance limits the tuning range quite severely, or requires a smaller inductance value that results in reduced output voltage swing (or a higher bias current). A reduced output voltage swing results in deterioration of the phase noise performance.

Using the inductor model implemented in the TSMC 0.25 μm design library, the equivalent parallel resistance versus inductance is calculated and shown in Fig. 27. The equivalent parallel resistance increases with increasing inductance, confirming the possibility of obtaining large output signal swing using the proposed architecture. Fig. 28 shows the tuning range simulation results. The designed 5.3 GHz VCO has 1.14 GHz (21.5%) tuning range. Typical structure VCO has also been designed using the same tank and power budget, and the tuning range is compared with the proposed structure in Fig. 28. Simulation data confirms the improved tuning range and reduced parasitic characteristics of the proposed architecture. The simulated phase noise results for both the proposed structure and the typical structure operating at 5 GHz are shown in Fig. 29. It has $-83.4\,\text{dBc/Hz}$ phase noise at 100 KHz offset and $-106\,\text{dBc/Hz}$ phase noise at 1 MHz offset from 5 GHz.

3.2. *20GHz VCO in BiCMOS*

A 20 GHz VCO with 5 GHz tuning range has been designed using the IBM 0.25μ m BiCMOS process. At the time of the initial design, we did not fully appreciate the

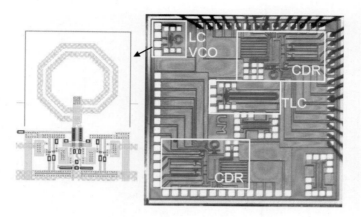

Fig. 30. Chip photograph and layout.

importance of the g_{m1}/g_{m2} ratio for R_{Eq} optimization. So unfortunately, in the experimental prototype design the g_{m1}/g_{m2} ratio (δ) is about 0.3 which is less than the optimal value. At 20GHz, the simulated $R_{Eq} \approx -500\,\Omega$ and $C_{Eq} \approx 13\,\text{fF}$. This includes the effect of the buffer. We note that C_{Eq} is about $1/9th$ the C_π of Q_1 which is equal to 126 fF. The designed VCO uses a two-stage emitter follower as buffers for each output. The capacitive impedance looking into this buffer works as a degeneration capacitance and contributes to the negative resistance as shown in [17]. But, as described in the previous section, the degeneration using active devices is more desirable and the buffer has been designed to have minimal impact on this design. Two back-to-back junction diodes described in Section 2.5 are used as the varactors, and the anode is connected to the base of the BJT to minimize the N/Substrate parasitic capacitance effect. The spiral inductor uses two turns to generate a 740 pH inductance. A single differential inductor is used for higher Q and to save area. This single inductor design without a center tap also prevents common mode oscillation. The chip microphotograph and VCO layout details are shown in Fig. 30.

For the measurements, we used RF probes to directly connect to the bare die. Single ended measurement setup has been used throughout. Fig. 31 shows the measured output power spectrum of the 20GHz oscillation signal. The measured tuning range is compared with the simulated tuning range in Fig. 32 (a). The measured oscillation frequency is slightly higher than the simulation results. This is because we overestimated the parasitic of the signal lines and probe pads. Measurement results show more than 5 GHz (25%) tuning range. This wide tuning was possible because of the low effective capacitance of the proposed negative resistance cell. Fig. 32 (b) shows simulated oscillation frequency at a fixed bias condition as a function of the temperature for different process corners. For these corner simulations, only the process corner parameter for bipolar devices and 6 sigma variations have been considered. Fig. 32 (b) shows that temperature and process variation results in about

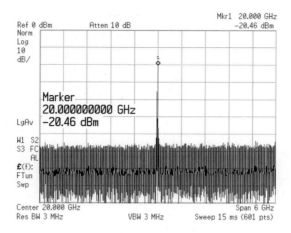

Fig. 31. Measured output spectrum.

4 GHz variation of the center frequency clearly showing the importance of a wide tuning range design that can tolerate temperature and process variations. The VCO core consumes only 2 mA, and each emitter follower branch drains 0.5 mA from a 4.5 V supply respectively. This low bias current compared to the other presented ∼ 20 GHz VCOs [8]- [11] supports the low-power design capability of the proposed topology. Despite the wide tuning range, the maximum output power variation was less than 3.5 dBm over the whole frequency tuning range as shown in Fig. 33, which is comparable to other presented works. A low power signal close to the noise floor is injected at the opposite side of differential output to remove the random drifting of the free running VCO as shown in Fig. 34 (a). The phase-noise of the injected signal was −121 dBc/Hz at 2 MHz offset from the 19.4 GHz carrier. At high injection levels, the phase noise of the VCO follows the phase noise of the injected signal source. But at low injection power levels, the locking range reduces and the phase noise approaches its intrinsic level as shown in Fig. 34 (b) [28]- [30]. Fig. 35 shows the measured injection locking range as a function of injected signal power and the measured phase noise at an injected signal power of -58dBm. At this injection level the lock range is about 800 KHz. The actual power delivered to the tank is much smaller because of the attenuation through the buffer stage. The phase noise is measured at a 2 MHz offset from 19.4 GHz to eliminate the phase noise attenuation through injection locking. The measured phase noise was −105.5 dBc/Hz. Simulations predicted a value of −110.6 dBc/Hz. This value is obtained using an NMOS noise coefficient (γ) of 2/3 and no induced gate noise contribution was included in the simulations. For short-channel devices, γ can reach as high as 2.5, and induced gate noise should be included. therefore our simulations underestimate the actual phase noise [31]. We believe the 5 dB discrepancy between measurement and simulation is from this non-perfect transistor noise model along with the noise in the control line and power supply that have not been included in the simulations.

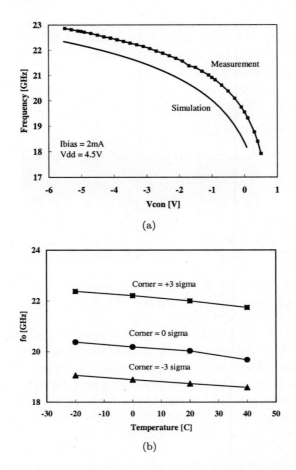

Fig. 32. Measured tuning range (a) and simulated frequency variation according to the temperature and process variation (b).

The figure of merit (FOM) defined in [6] has been used widely to compare the performance of VCOs, but it doesn't include the impact of the frequency tuning range. Recently a new figure of merit (FOM_T) including the frequency tuning range has been presented in [7]. Both FOM and FOM_T have been used to compare our design with other VCOs. The results of this comparison with other VCOs working around 20 GHz is shown in Table I. Among the circuits that have been compared, this work shows the highest FOM and FOM_T, and supports the effectiveness claim of the proposed topology for low-power low-noise, high-frequency VCO design.

4. Summary and Conclusions

We initially analyzed the maximum attainable oscillation frequency of VCOs that utilizes the widely used cross-coupled negative-g_m cell. This analysis showed that

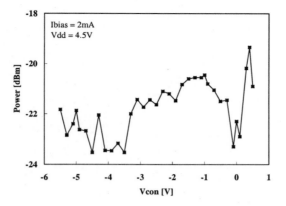

Fig. 33. Measured power variation.

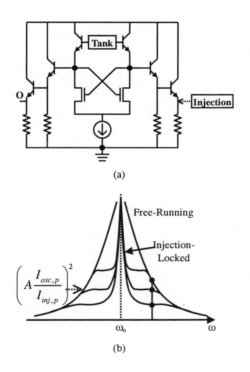

Fig. 34. Phase noise measurement setup.

the finite base/gate resistance of the transistor and the equivalent parasitic capacitance (C_{Eq}) of a negative resistance cell limit the maximum attainable oscillation frequency. The following analysis on the integrated differential negative resistance cell using a capacitive degeneration technique proves that it has a much higher max-

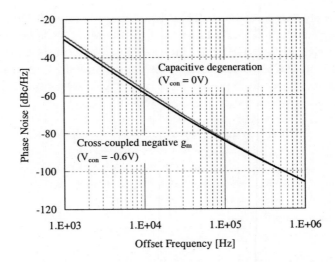

Fig. 35. Locking range and phase noise measurement results.

Table 1.   Comparison of figure of merits.

Reference	Technology	Tuning Range (%)	f_0 (GHz)	*FOM*	FOM_T
[8]	SiGe	14.5	20	−163.0	−166.3
[9]	SiGe	3	20	−154.7	−144.2
[10]	InP	36	21.5	−154.3	−165.4
[11]	SiGe	15	20	−170.0	−173.6
This work	SiGe	25	20	−172.7	−180.6

imum attainable oscillation frequency and a much smaller equivalent shunt capacitance as compared to the cross-coupled negative-g_m cell. This makes the VCO using a capacitively degenerated negative resistance cell a good choice for high-frequency integrated oscillators working close to the process f_T. The small equivalent capacitance is also useful to increase the tuning range of the oscillators operating at relatively low frequencies. The example design of a 5.3 GHz VCO in 0.25 μm CMOS technology shows a wider tuning range than VCO using a cross-coupled negative-g_m cell without compromising on phase noise or power consumption.

Unfortunately, the capacitive degeneration technique reduces the effective transconductance of the cell, and hence the effective negative resistance, R_{Eq}, that is generated. We presented an active capacitive degeneration topology that can alleviate this problem. It uses a cross-coupled transistor pair as a degeneration cell. The cross-coupled transistor pair contributes additional conductance, and allows for the negative resistance cell to maintain a high maximum attainable oscillation frequency and have better negative resistance characteristics in comparison to the simple capacitively degenerated cell. These properties combined with its small effective capacitance enables low-power low-noise high-frequency VCO implementations. A prototype design for a 20 GHz fully integrated *LC* VCO implementation in the IBM SiGe 0.25 μm BiCMOS technology was demonstrated. A comparison of the figure of merit to previously reported 20 GHz VCOs supports the effectiveness of the active capacitive degeneration topology.

Although the VCOs using a capacitively degenerated negative resistance cell has been around in discrete design, the study for its application for fully integrated differential VCOs are fairly recently. The work presented in this paper proves the effectiveness of the capacitive degeneration technique through analytical derivations, simulations and measurements. The desirable characteristics discussed in this paper make the capacitive degeneration technique a promising candidate for high-frequency integrated oscillators.

Acknowledgments

The authors would like to thank the Semiconductor Research Corporation and other sponsoring companies of the "2002-2003 SiGe BiCMOS Design Challenge" for chip fabrication and tool support. They would also like to thank K. Johnson of Honeywell and P. Cheung of Bermai for help with measurements and A. Gopinath, J. Kim, Shubha B., and Y. Tseng of the University of Minnesota for valuable discussions.

References

1. Donhee Ham and Ali Hajimiri, "Concepts and Methods in Optimization of Integrated *LC* VCOs," *IEEE Journal Solid-State Circuits,* vol. 36, pp. 896-909, June 2001.
2. Ali Hajimiri and Thomas H. Lee, "Design Issues in CMOS Differential *LC* Oscillators," *IEEE Journal Solid-State Circuits,* vol. 34, pp. 717-724, May 1999.
3. Emad Hegazi, Henrik Sjöland, and Asad A. Abidi, "A Filtering Technique to Lower *LC* Oscillator Phase Noise," *IEEE Journal Solid-State Circuits,* vol. 36, pp. 1921-1930, Dec. 2001.
4. Markus Zannoth, Bernd Kolb, Joseph Fenk, and Robert Weigel, "A Fully Integrated VCO at 2 GHz," *IEEE Journal Solid-State Circuits,* vol. 33, pp. 1987-1991, Dec. 1998.
5. Nikolay T. Tchamov, Tero Niemi, and Niko Mikkola, "High-performance differential VCO based on Armstrong oscillator topology," *IEEE Journal Solid-State Circuits,* vol. 36, pp. 139-141, Jan. 2001.
6. Jean-Olivier Plouchart, Herschel Ainspan, Mehmet Soyuer, and Albert Ruehli, "A Fully-Monolithic SiGe Differential Voltage-Controlled Oscillator For 5GHz Wireless Applications," *IEEE RFIC Symp. Dig.,* 2000, pp. 57-60.
7. Jonghae Kim, Jean-Olivier Plouchart, Noah Zamdmer, Melanie Sherony, Yue Tan, Meeyoung Yoon, Robert Trzcinski, Mohamed Talbi, John Safran, Asit Ray, and Lawrence Wagner, "A Power-Optimized Widely-Tunable 5-GHz Monolithic VCO in a Digital SOI CMOS Technology on High Resistivity Substrate," *Int. Symp. Low Power Electronics and Design,* Aug. 2003, pp. 434-439.
8. P. Deixler, R. Colclaser, D. Bower, N. Bell, W. De Boer, D. Szmyd, S. Bardy, W. Wilbanks, P. Barre, M. v. Houdt, J. C. J. Paasschens, H. Veenstra, E. v. d. Heijden, J. J. T. M. Donkers, and J. W. Slotboom, "QUBiC4G: A f_T/f_{max}=70/100GHz 0.25μm Low

Power SiGe-BiCMOS Production Technology with High Quality Passives for 12.5Gb/s Optical Networking and Emerging Wireless Applications up to 20GHz," *IEEE Proc. BCTM,* 2002, pp. 201-204.

9. K. Ettinger, A. Stelzer, C. G. Diskus, W. Thomann, J. Fenk, and R. Weigel, "Single-Chip 20-GHz VCO and Frequency Divider in SiGe Technology," *IEEE MTT-S Int. Microwave Symp. Dig.,* 2002, pp. 835-838.

10. Hormoz Djahanshahi, Namdar Saniei, Sorin P. Voinigescu, Michael C. Maliepaard, and C. André T. Salama, "A 20-GHz InP-HBT Voltage-Controlled Oscillator With Wide Frequency Tuning Range," *IEEE Trans. Microwave Theory Tech.,* vol. 49, pp. 1566-1571, Sept. 2001.

11. S. P. Voinigescu, D. Marchesan, and M. A. Copeland, "A Family of Monolithic Inductor-Varactor SiGe-HBT VCOs For 20GHz To 30GHz LMDS And Fiber-Optic Receiver Applications," *IEEE RFIC Symp. Dig.,* 2000, pp. 173-176.

12. Hugo Veenstra and Edwin van der Heijden, "A 19-23 GHz Integrated LC-VCO in a production 70 GHz F_T SiGe technology," *ESSCIRC,* Sept. 2003, pp. 349-352.

13. Robert T. Oyafuso, "An $8-18$ GHz FET YIG-Tuned Oscillator," *IEEE Int. Microwave Symp. Dig.,* 1979, pp. 183-184.

14. G. D. Vendelin, A. M. Pavio, and U. L. Rhode, *Microwave Circuit Design Using Linear and Non-Linear Techniques.* New York: Wiley, ch. 6, 1990.

15. Mehmet Soyuer, Joachim N. Burghartz, Herschel A. Ainspan, Keith A. Jenkins, Peter Xiao, Arvin R. Shahani, Margaret S. Dolan, and David L. Harame, "An $11-$GHz $3-$V SiGe Voltage Controlled Oscillator with Integrated Resonator," *IEEE Journal Solid-State Circuits,* vol. 32, pp. 1451-1454, Sept. 1997.

16. A. P. S. Khanna, Ed Topacio, Ed Gane, and Danny Elad, "Low Jitter Silicon Bipolar Based VCOs for Applications in High Speed Optical Communication Systems," *IEEE Int. Microwave Symp. Dig.,* 2001, pp. 1567-1570.

17. Hugo Veenstra and Edwin van der Heijden, "A 35.2-37.6GHz LC VCO in a 70/100GHz f_T/f_{max} SiGe Technology," *ISSCC Dig. Tech. Papers,* 2004, pp. 394-395.

18. Byunghoo Jung and Ramesh Harjani, "A Wide Tuning Range VCO Using Capacitive Source Degeneration," *IEEE Int. Symp. on Circuits and Systems,* May 2004, pp. 145-148.

19. B. Jung and R. Harjani, "A 20GHz VCO with 5GHz Tuning Range in 0.25μm SiGe BiCMOS," *ISSCC Dig. Tech. Papers,* 2004, pp. 178-179.

20. Byunghoo Jung and Ramesh Harjani, "High-Frequency *LC* VCO Design Using Capacitive Degeneration," *IEEE Journal Solid-State Circuits,* vol. 39, pp. -, Dec. 2004.

21. Jing-Hong C. Zhan, Jon S. Duster, and Kevin T. Kornegay, "A $25-$GHz Emitter Degenerated *LC* VCO," *IEEE Journal Solid-State Circuits,* vol. 39, pp. 2062-2064, Nov. 2004.

22. Behzad Razavi, *RF Microelectronics.* NJ: Prentice Hall PTR, 1998.

23. Behzad Razavi, "A Study of Noise in CMOS Oscillators," *IEEE Journal Solid-State Circuits,* vol. 31, pp. 331-343, Mar. 1996.

24. Behzad Razavi, "A 1.8GHz CMOS Voltage-Controlled Oscillator," *ISSCC Dig. Tech. Papers,* 1997, pp. 388-389.

25. John R. Long and Miles A. Copeland, "The Modeling, Characterization, and Design of Monolithic Inductors for Silicon RF IC's," *IEEE Journal Solid-State Circuits,* vol. 32, pp. 357-369, Mar. 1997.

26. John W. M. Rogers, José A. Macedo, and Calvin Plett, "The Effect of Varactor Nonlinearity on the Phase Noise of Completely Integrated VCOs," *IEEE Journal Solid-State Circuits,* vol. 35, pp. 1360-1367, Sept. 2000.

27. C. Patrick Yue, S. Simon Wong, "On-Chip Spiral Inductors with Patterned Ground

Shields for Si-Based RF IC's," *IEEE Journal Solid-State Circuits,* vol. 33, pp. 743-752, May 1998.

28. Robert Adler, "A Study of Locking Phenomena in Oscillators," *Proc. IRE and Waves and Electronics,* vol. 34, no. 6, pp. 351-357, June 1946.

29. Behzad Razavi, "A Study of Injection Pulling and Locking in Oscillators," *IEEE Custom Integrated Circuits Conf.,* 2003, pp. 305-312.

30. Xiangdong Zhang, Brian J. Rizzi, and James Kramer, "A New Measurement Approach for Phase Noise at Close-in Offset Frequencies of Free-Running Oscillators," *IEEE Trans. Microwave Theory Tech.,* vol. 44, pp. 2711-2717, Dec. 1996.

31. Behzad Razavi, "Prospects of CMOS Technology for High-Speed Optical Communication Circuits," *IEEE Journal Solid-State Circuits,* vol. 37, pp. 1135-1145, Sept. 2002.

International Journal of High Speed Electronics and Systems
Vol. 15, No. 2 (2005) 353–375
© World Scientific Publishing Company

FULLY INTEGRATED FREQUENCY SYNTHESIZERS:
A TUTORIAL

Sung Tae Moon

Electrical Engineering, Texas A&M University
College Station, TX 77840, United States of America
stmoon@tamu.edu

Ari Yakov Valero-López

Agere Systems, Storage Division, Read Channel Development
Allentown, PA 18109, United States of America
av4@agere.com

Edgar Sánchez-Sinencio

Electrical Engineering, Texas A&M University
College Station, TX 77840, United States of America
sanchez@ee.tamu.edu

Frequency synthesizer is a key building block of fully-integrated wireless communications systems. Design of a frequency synthesizer (FS) requires the understanding of not only the circuit-level but also of the transceiver system-level considerations. The FS design challenge involves strong trade-offs, and often conflicting requirements. In this tutorial, the general implementation issues and recent developments of frequency synthesizer design are discussed. Simplified design approach should provide readers with sufficient intuition for fast design and troubleshooting capability. Open problems in this FS field are briefly discussed.

Keywords: Analog integrated circuits; CMOS/BiCMOS RF; frequency synthesizer; phase-locked loop

1. Introduction

Wireless communications gained popularity as the electronics industry introduced accessible consumer products leading the emerging market. The most effective way to save production cost and to minimize form factor has been the monolithic implementation of the entire RF transceiver on a single chip. In a fully integrated system, the frequency synthesizer design represents a major challenge since the circuit has to meet stringent and conflicting requirements.

A frequency synthesizer (FS) is a device capable of generating a set of signals of given output frequencies with very high accuracy and precision from a single reference frequency. The signal generated at the output of the frequency synthesizer is commonly known as local oscillator (LO) signal, since it is used in communication systems as the reference oscillator for frequency translation as shown in Fig. 1. The reference signal at high frequency is used to downconvert the incoming signal into a lower frequency where it can be processed to extract the information it is carrying. The same reference signal can be used to upconvert a desired message to an RF frequency, such that it can be transmitted over the medium.

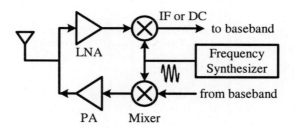

Fig. 1. The role of a frequency synthesizer in a communication transceiver

Normally, the FS output signal is a sinusoidal tone plus harmonic tones that are added due to non-linearities. Fundamentally, the whole frequency synthesizer system is designed to ensure the accuracy of its output frequency under any condition. In fact, the accuracy requirements are so tight that the accuracy are in the order of tens of ppm. For example, the final frequency accuracy in Wireless LAN 802.11a standard is 20 ppm, which translates into 116 kHz for a carrier frequency of 5.805 GHz.[1] In addition to the frequency accuracy, the spectral purity of the output signal and the settling time determine the performance merits of a frequency synthesizer.

This tutorial discusses the general design considerations and recent developments of frequency synthesizers design. Section 2 studies a methodology of interpreting communication standards into circuit specifications as a top-down design strategy. Section 3 and 4 covers the details of conventional frequency synthesizer. Section 5 summarizes the recent development of advanced techniques to improve the performance of frequency synthesizers. The tutorial assumes the reader has a basic understanding of PLL operation[2] and builds on that knowledge to describe more detailed design issues particular to wireless communications frequency synthesizers.

2. Interpreting Specifications

The detailed specifications for the transistor-level design of frequency synthesizers are not readily available from the standard, but are embedded within the description of the requirements for the communication system. Also, particular characteristics of the system design set constrains in the specifications of the frequency synthesizer. For example, even though the RF frequencies are set for a given standard, the selection of a given intermediate frequency (IF) determines the required output frequency range of the synthesizer. Table 1 is used to illustrate the information in some standard documents, that is relevant to frequency synthesizer design. Full details of several wireless communication standards can be found in the literature.[1,3,4,5]

Table 1. Short range wireless communications standards

	Bluetooth	802.11a	802.11b	802.11g
Bit rate	1 Mbps	54 Mbps	11 Mbps	54 Mbps
Sensitivity	-70 dBm	-82 dBm	-76 dBm	-76 dBm
Frame Error Rate	10^{-3} (BER)	10^{-5}	8×10^{-2}	8×10^{-2}
Band (MHz)	2400–2479	5180–5805	2412–2472	2412–2472
Channel Spacing	1 MHz	20 MHz	5 MHz	20 MHz
Accuracy	±75 kHz[a]	±20 ppm	±25 ppm	±25 ppm
Settling	< 259 μs	224 μs	224 μs	224 μs
Interference	+40 dB at 3 MHz	+32 dB at 40 MHz	+35 dB at 25 MHz	+35 dB at 25 MHz

[a]Equivalent to 30 ppm

2.1. *Frequency Band and Tuning Range*

Every communication standard utilizes a specific frequency band in the spectrum of electro-magnetic waves according to the usage models, and the regulations of the governing body. For instance, the 2.4 GHz Industrial-Scientific-Medical (ISM) band is most popular for short range communication standards such as Bluetooth and Wireless LAN, because the usage of the ISM band is free and the frequency is high enough to limit the reach of the transmitted signal.

In phase-locked loop (PLL) based frequency synthesizers, the tuning range of the voltage controlled oscillator (VCO) determines the limits on the overall system tuning range. The tuning range of the VCO should be much larger than the frequency band of interest since it will have large range of uncertainty due to process variations and modelling uncertainties. A 20% deviation in either inductance or capacitance in a LC oscillator will result in more then 10% error in the output frequency.

Other factors, such as the linear range at the output voltage of the charge pump (CP) can further limit the tuning range of the synthesizer. The design should account for the limits of both, the VCO control voltage and CP output linear range to ensure the synthesizer can operate properly. If the CP cannot provide the designed current amplitude for certain output voltages, the system transfer function loses its gain and may become unstable. The voltage swing can be severely limited if the charge pump has a cascode output stage for improved output impedance.[6]

2.2. *Channel Agility and Settling Time*

Whenever the transmission or reception channel switches in a communication system, the transceiver must change its local oscillator frequency to synchronize with the received/transmitted signal. Since most frequency synthesizers utilize a feedback mechanism to control the accuracy of the output frequency, and minimize the difference between the output and the target frequency, the switching of the output frequency cannot be instantaneous. The output frequency approximately follows the step response of a second order system for very small phase errors. This condition holds only when the frequency step is much smaller than the center frequency, as in narrow-band systems. For large frequency steps in wide-band systems, the response will slow down due to very non-linear behavior associated with large phase errors.

For instance, in the Bluetooth standard, the frequency synthesizer settling time is not clearly defined, but it can be calculated from the relationship between the time slot length and the packet length. Since the Bluetooth standard uses frequency hopping at 1600 hops per second, the transceiver is only allowed to transmit within a time slot of $T_{slot} = 625$ μs. The length of a standard single packet to be transmitted in a time slot is 366 bit long, corresponding to $T_{pkt} = 366$ μs. Thus the downtime between two consecutive time slots is,

$$T_{down} = T_{slot} - T_{pkt} = 259 \text{ } \mu s \tag{1}$$

The transceiver must complete a transition between transmitting and receiving during the T_{down} period, including the settling of the frequency synthesizer. Note that the settling time of the frequency synthesizer is only a fraction of the turnaround time because the blocks following the mixer, such as variable gain amplifier (VGA), also need certain amount of time to settle once the frequency synthesizer is settled.

Wireless LAN standards explicitly specify channel agility to be 224 μs in the standard section 18.4.6.12. A frequency synthesizer is considered to be settled when the center frequency is stable within the frequency accuracy limit, which is ± 60 kHz for the case of 802.11b.

2.3. *Spectral Purity*

The spectral purity of the local oscillator is usually not explicitly specified in most of the communication standards. Instead, phase noise and spurious signal specifications are usually derived from adjacent channel interference requirements.[6] The strongest adjacent channel interferences of several popular short-range standards are listed in Table 1.

The effect of phase noise and adjacent channel interference is shown in Fig. 2. While the signal (P_{Sig}) is downconverted to DC or IF by the LO signal (P_{LO}), the interference (P_{Int}) is also downconverted to DC or IF by the phase noise (P_N) and is added to the signal of interest. Since the phase noise is a random process, the effective bandwidth (P_{BW}) is added to calculate the total power. The signal to noise ratio (SNR) of the baseband signal is the difference of the power of the two, and it must be larger than the minimum SNR required to meet the receiver bit error rate (BER) requirement.

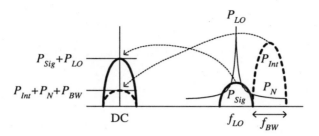

Fig. 2. The effect of phase noise and interference

$$SNR = (P_{Sig} + P_{LO}) - (P_{Int} + P_N + P_{BW}) > SNR_{min} \qquad (2)$$

After rearrangement,

$$P_N - P_{LO} < (P_{Sig} - P_{Int}) - P_{BW} - SNR_{min} \qquad (3)$$

where $P_N - P_{LO}$ denotes the phase noise requirement in dBc – a power spectrum density relative to the carrier power. For example, from Table 1, Bluetooth standard specifies an interferer of +40 dB at 3 MHz away from the desired signal. The channel bandwidth is 1 MHz, which translates into $P_{BW} = 10 \log 10^6 = 60$ dB. The minimum SNR requirement for a BER of 10^{-3} is 18 dB, which can be determined from system level baseband simulations.[a] Substituting these numbers in Eq. (3), the phase noise requirement is -118 dBc at 3 MHz from carrier. This calculation assumes the phase noise is white within the channel bandwidth. A realistic design goal should include some margin from the calculated value.

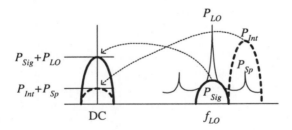

Fig. 3. The effect of reference spur and interference

Reference spur can be a especially serious problem if the system uses narrow channel spacing and the spur coincides with the adjacent channels as shown in Fig. 3. This kind of situation can happen when implementing Bluetooth transceivers with an integer-N type frequency synthesizer. The calculation is similar to the one

---

[a]For this tutorial, SytemView^{TM} software is used to simulate GFSK coded baseband signal for Bluetooth system. The BER of the final signal is measured while sweeping the additional noise power.

previously presented for phase noise case except that the interference is downconverted by spurious signal, which is considered as a single tone. With the SNR of the baseband signal being,

$$SNR = (P_{Sig} + P_{LO}) - (P_{Int} + P_{Sp}) > SNR_{min} \qquad (4)$$

After rearrangement,

$$P_{Sp} - P_{LO} < P_{Sig} - P_{Int} - SNR_{min} \qquad (5)$$

where $P_{Sp} - P_{LO}$ denotes the power of spurious signal in dBc, relative to the carrier power. For example, Bluetooth standard specifies an interferer of +30 dB at 2 MHz away from the desired signal. The reference spur can be also at 2 MHz away from the carrier if the frequency of the reference signal is 2 MHz. The minimum SNR requirement is 18 dB, same as the previous example. Substituting the numbers in Eq. (5), the spurious signal requirement results in −48 dBc at 2 MHz from carrier.

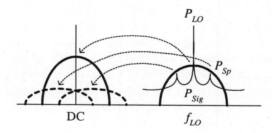

Fig. 4. The effect of reference spur in 802.11b system

In the case of Wireless LAN 802.11b as shown in Fig. 4, the reference spur can fall *within* the received signal, but not in the adjacent channel because the channel bandwidth can be larger than the reference frequency. System level simulations are required to determine the specific level of spur that degrades the receiver BER below the given specification.[b] Simulation results are presented in Fig. 5. The SNR of the input signal swept from 10.5 dB to 14 dB, while four different spur power of −34, −28, −22, and −16 dB are degrading the input signal. The result shows that the reference spur must be at least 25 dB below the carrier signal to keep a BER better than 10^{-5} when the input SNR is 12 dB. This requirement also needs

---

[b]The CCK coded baseband signal of 802.11b system is simulated using SytemView^{TM} software. The baseband signal is up-converted by 2 MHz and then added to the original baseband signal. The degradation of the final signal is measured in terms of BER.

additional margin for a realistic design.

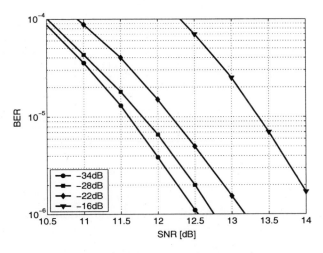

Fig. 5. The effect of reference spur at 2 MHz in 802.11b system

Table 2 summarizes the mapping relationship between the communication standard and the building block specification. It is possible for several aspects of the standard to be mapped into a single specification, and vice versa. For illustration purpose, a specific example for 802.11b standard is given in separate columns.

Table 2. Summary of specification mapping

Standard		Specification	
General	802.11b	General	802.11b
Frequency Band	2412–2472 MHz	Tuning range	2412–2472 MHz
Channel spacing	5 MHz	Tuning step	1 MHz
Hopping rate	N/A	Settling time	224 μs
Packet structure	N/A	Settling time	N/A
Interference	+35 dB at 25 MHz	Phase noise	-126 dBc at 25 MHz
Interference	N/A	Spur rejection	-25 dBc at 2 MHz

3. Types of Frequency Synthesizers

3.1. *PLL based Integer-N Synthesizer*

The most popular technique of frequency synthesis is based on the use of a phase-locked loop (PLL). The loop is synchronized or locked when the phase of the input signal and the phase of the output from the frequency divider are aligned. As shown in Fig. 6, the output of the VCO in the integer-N synthesizer is divided and phase-locked to a stable reference signal. Once the loop is locked, the output frequency equals the reference frequency times N.

$$f_{out} = N \cdot f_{REF} \tag{6}$$

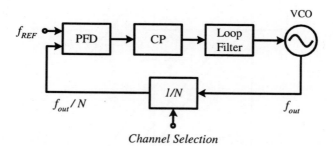

Fig. 6. Integer-N architecture

Integer-N architecture is the preferred solution for minimizing power consumption and die area due to its simplicity. The integer-N architecture, however, lacks the flexibility of arbitrarily choosing f_{REF} as is possible in more complex architectures. Since f_{REF} is fixed by channel spacing requirements, the loop bandwidth can be severely limited, especially since it has to be significantly lower than f_{REF} for stability considerations.

Although, the integer-N synthesizers can generate output frequencies in steps of f_{REF}, the channel spacing is not necessarily equal to f_{REF}. The maximum possible f_{REF} can be calculated as follows: First, the channel frequencies must be integer multiples of f_{REF} as shown in Eq. (6), but at the same time the channel spacing also has to be an integer multiple of f_{REF}. To satisfy both conditions, the f_{REF} has to be the greatest common divisor (GCD) of the channel frequency and the channel spacing. For example, Wireless LAN 802.11b standard specifies channels from 2412 MHz to 2472 MHz in steps of 5 MHz. Thus, the maximum possible f_{REF} is $GCD(2412 \text{ MHz}, 5 \text{ MHz}) = 1$ MHz. For a different example, Wireless LAN 802.11a standard specifies a channel at 5805 MHz and a step of 20 MHz. In this case, the maximum possible f_{REF} is $GCD(5805 \text{ MHz}, 20 \text{ MHz}) = 5$ MHz.

3.2. PLL based Fractional-N Synthesizer

An inherent shortcoming of the integer-N synthesizer is the limited option for the reference frequency, f_{REF}, because of the integer-only multiplication. A fractional-N synthesizer architecture solves this problem by allowing fractional feedback ratios. Shown in Fig. 7, the fractional-N synthesizer has a dual modulus divider that can switch its division ratio between N and $N + 1$. By dividing the VCO frequency by N during K VCO cycles and $N + 1$ during $(2^k - K)$ VCO cycles, it is possible to make the average division ratio equal to $N + K/2^k$, assuming a k bits accumulator controlling the prescaler. Thus,

$$f_{out} = (N + \alpha) \cdot f_{REF} \quad \text{,where} \ \ 0 < \alpha < 1 \tag{7}$$

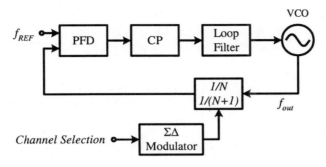

Fig. 7. Fractional-N architecture

However, if the division modulus is switched periodically, the output is modulated by the beat frequency of the fractional modulus. It can be shown that the output spectrum has tones at αf_{REF}, $2\alpha f_{REF}$ and so on, relative to the carrier frequency. These are fractional spurs and can be problematic since they are very close to the carrier.

The fractional spurs can be reduced by breaking the regularity of the division modulus switching period, effectively making the beat frequency randomized. A dithering mechanism using $\Sigma\Delta$ modulator can not only randomize the beat frequency, but shape the noise spectrum so that it has more power at higher frequency. The high frequency quantization noise is filtered by the loop filter of the PLL. A combination of the order of the $\Sigma\Delta$ modulator and loop filter order can reduce the high frequency quantization noise at levels that make the effect of the noise negligible.[7]

3.3. *Direct Digital Synthesizer (DDS)*

A fundamental reason that a feedback control loop is used in the implementation of frequency synthesizers is because the relationship between the control voltage and the output frequency of a VCO is unpredictable and subject to variations from unwanted excitations. If a VCO's output signal frequency were always predictable with no variation, there would be no need to use feedback control to correct the error in frequency. The output of the VCO would be used directly as the final output of the frequency synthesizer. In this hypothetical system, there would be no problem of stability and settling time. The settling time would be only limited by the gate delay of the channel selection input.

DDS generates its output signal from the digital domain and converts it in analog waveform through a digital-to-analog converter (DAC) and filtering as shown in Fig. 8. Since the waveform is directly shaped from the amplitude values from a read-only-memory (ROM), it doesn't require feedback and it has all the advantages of the hypothetical system previously described. In addition, it has other advantages such as low phase noise and possibility of direct digital modulation. The DDS is a suitable choice when the carrier frequency has to be settled very fast with very low phase noise.[8] The application of the FS is to generate frequency-hopped carrier signals for NMT-900 cell phone standard. Another usage of DDS is when extremely fine frequency resolution is required.[9] This synthesizer covers a bandwidth from DC to 75 MHz in steps of 0.035 Hz with a switching speed of 6.7 ns.

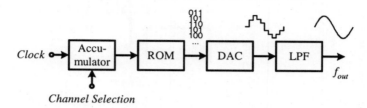

Fig. 8. Direct digital synthesizer block diagram

The most serious shortcoming of DDS is speed: the clock of the digital circuitry has to be at least twice as high as the output frequency. Operating a ROM and a DAC at 4.8 GHz to generate 2.4 GHz output signals can be challenging in current technologies, if at all possible, and power consumption will be prohibitively high. In addition, large quantization noise and harmonic distortion of high speed DACs can degrade the spectral purity of the output signal. Using an analog mixer to upconvert a low frequency synthesized signal, in order to generate high frequency outputs without an excessively high frequency clock, has been reported in literature.[10] However, it is a costly solution since it needs an extra analog PLL and high frequency mixers.

4. Phase Locked Loop (PLL) Design

This section covers the fundamentals of PLL design for frequency synthesizers. Rather than focusing on circuit implementation issues, system level designs such as loop transfer function and stability considerations are addressed with insightful observations. Extensive PLL design techniques can be found in the literature.[2,11,12]

4.1. *Charge-pump PLL*

Virtually all of the PLL-based frequency synthesizers utilize a charge-pump PLL that was first introduced by Gardner.[13] Charge-pump PLL has important advantages that make it suitable for the implementation of frequency synthesizers. These advantages include:

(i) The operation of phase frequency detector (PFD) makes the frequency acquisition range not limited by loop bandwidth but only by VCO tuning range.

(ii) Due to poles at the origin, charge-pump PLL has infinite open-loop gain at DC, which make the static phase error to be ideally zero.

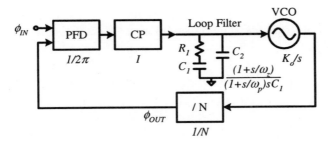

Fig. 9. Charge-pump PLL block diagram and linear approximation

A simplified block diagram of the charge-pump (CP) PLL is shown in Fig. 9. The fundamental process of operation is as follows. First, the VCO oscillates at its natural frequency assuming the control voltage is arbitrary at the beginning. The PFD compares the phase difference between the reference signal ϕ_{IN} and the VCO output divided by the frequency divider, ϕ_{OUT}. The output of the PFD is a series of pulses whose duty cycle is proportional to the phase difference $\phi_{IN} - \phi_{OUT}$. The CP converts the voltage pulses into current pulses with a predetermined amplitude I. The loop filter converts the current pulses into a low-pass filtered voltage signal that controls the frequency of the VCO. If the feedback is negative, the error between ϕ_{IN} and ϕ_{OUT} gradually become smaller and smaller until $\phi_{IN} = \phi_{OUT}$. In this state the loop is referred to be *locked*. Once the loop is locked, the frequency of the VCO output is equal to the frequency of the reference multiplied by the feedback factor N.

The process of locking is not instantaneous because the loop has a limited bandwidth. The transfer function of the loop has to be studied to estimate the behavior of the loop during its transient operation. Since the operation of the PFD and CP is performed in the discrete-time domain, the complete transfer function becomes complicated due to the z-transform representation. A more intuitive equation can be obtained by assuming the phase error is small. With this assumption, the PFD and CP are modelled as simple gain blocks, $1/2\pi$ and I respectively, as shown in Fig. 9.

The linear approximation gives two critical equations useful for the initial design of a PLL. The first equation is an open-loop transfer function which is ϕ_{OUT}/ϕ_{IN} assuming the loop is opened between the frequency divider and the PFD.

$$H_{open}(s) = \frac{\phi_{OUT}}{\phi_{IN}} = \frac{K_D K_o (1 + s/\omega_z)}{(1 + s/\omega_p)s^2} \tag{8}$$

where $K_D = I/(2\pi C_1 N)$, $\omega_z = 1/(R_1 C_1)$ and $\omega_p \simeq 1/(R_1 C_2)$. The open-loop transfer function is important because its phase margin indicates how stable the system will be after the loop is closed. Note that there are two poles at the origin and a stabilizing zero is required to compensate for them. Details of PLL stability are covered in section 4.2.

The second equation is a closed-loop transfer function ϕ_{IN}/ϕ_{OUT}. It can also be calculated from $H_{open}(s)/(1 + H_{open}(s))$.

$$H_{closed}(s) = \frac{\phi_{OUT}}{\phi_{IN}} = \frac{1 + s/\omega_z}{1 + s/\omega_z + s^2/(K_D K_o) + s^3/(\omega_p K_D K_o)} \tag{9}$$

For simplicity, it is assumed that ω_p is placed at very high frequency with respect to the natural frequency $\omega_n = \sqrt{K_D K_o}$, then the transfer function becomes second order.

$$H'_{closed}(s) \simeq \frac{1 + s/\omega_z}{1 + s/\omega_z + s^2/(K_D K_o)} \tag{10}$$

The step response of the closed-loop transfer function shows the locking transient, and settling time performance can be determined from the transient waveform. Analytic solution of the settling time can be derived from the second order transfer function. The details of the settling analysis is covered in section 4.3.

4.2. *Stability*

As in any feedback system, stability is one of the most important aspects of the design considerations of frequency synthesizers. A potentially unstable synthesizer will generate an output signal whose frequency does not converge but *oscillates* between certain frequency limits. The unstable output signal appears similar to narrow-band FM modulated signal.

There are two sources for the stability limit in charge-pump PLL. The first comes from the fact that the operation of PFD and CP is in the discrete-time domain. Loop bandwidth has to be carefully chosen so that the linear approximation is not violated i.e. $\omega_c < \omega_{REF}$. The second comes from the two poles at the origin in the open-loop transfer function. A stabilizing zero can compensate for the effect of the double poles at crossover frequency. More detailed analysis on stability limit follows.

First, the charge-pump PLL has a critical stability limitation due to the discrete nature of the PFD and CP output. The PLL operates as a sampled system and not as a straightforward continuous-time circuit. It is known that a sampled second-order PLL will become unstable if the loop gain is made so large that the bandwidth becomes comparable to the sampling frequency. Limited loop gain sets upper boundary of the loop bandwidth obtainable for a given input reference frequency. Gardner's stability limit states that:[13]

$$\omega_n^2 < \frac{\omega_{REF}^2}{\pi(\pi + \omega_{REF}/\omega_z)} \tag{11}$$

The relationship between the natural frequency (ω_n) and the loop bandwidth (ω_c) is approximately:

$$\omega_c \simeq \omega_n^2/\omega_z \tag{12}$$

for critically damped and overdamped system. Substituting Eq. (12) into Eq. (11), it can be rewritten as:

$$\omega_c < \frac{\omega_{REF}}{\pi(1 + \pi\omega_z/\omega_{REF})} \tag{13}$$

which indicates that the loop bandwidth (ω_c) has to be significantly lower than the frequency of the reference input signal (ω_{REF}). Commonly ω_c is chosen below one-tenth of ω_{REF} to guarantee stability. Another important factor to consider when determining the loop bandwidth is the size of the capacitors to realize the bandwidth. If the loop bandwidth is too narrow, the size of the capacitors can be excessively large to be implemented in a fully-integrated solution. Using dual-pass active filter[14] or impedance multiplier[15] are proposed to emulate a large capacitance without consuming huge die area. Their application is limited to a multiplication factor no more than 20 due to uncertainties from mismatch. Furthermore, the additional active device in the signal path can degrade phase noise and increase reference spurs due to leakage current.

The second stability limit comes from the open-loop transfer function. As has already been shown in Eq. (8), the open-loop transfer function of a charge-pump PLL has two poles at the origin, which makes the loop inherently unstable. A zero should be placed at a lower frequency than the crossover frequency to make the phase margin large enough ($> 45°$). Since the zero reduces the slope of the magnitude response, an additional pole at a higher frequency than the crossover frequency is also required to maintain adequate spurious signal rejection.

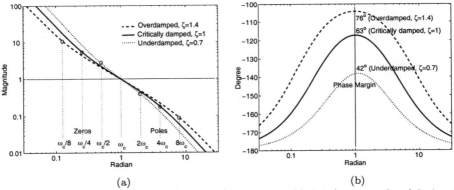

(a)   (b)

Fig. 10. The effect of pole/zero placement on phase margin (a) Pole/zero are placed 2, 4 and 8 times crossover frequency ω_c. (b) Phase margin increases from 42° to 76° as pole/zero are placed farther apart.

Three examples of different pole/zero placements are shown in Fig. 10. The open-loop transfer functions of those three examples referred to Eq. (8) are

$$H_{under}(s) = \frac{1 + 2s}{2(1 + s/2)s^2} \tag{14}$$

$$H_{critical}(s) = \frac{1 + 4s}{4(1 + s/4)s^2} \tag{15}$$

$$H_{over}(s) = \frac{1 + 8s}{8(1 + s/8)s^2} \tag{16}$$

When a zero is located at $1/4$ of the crossover frequency (w_c) and a pole is placed at 4 times of w_c, the loop is critically damped with the damping ratio of 1. A phase margin of $63°$ can be achieved. When the zero is at $w_c/2$ and the pole is at $2w_c$, the loop is underdamped with the damping ratio of 0.707. With an underdamped loop, the phase margin is lowered to $42°$ and the transient signal overshoots. When the zero is at $w_c/8$ and the pole is at $8w_c$, the loop is overdamped with a damping ratio of 1.414. With an overdamped loop, the phase margin is increased to $76°$ but the settling time is degraded due to slow response.

Normally a critically damped loop works best for a typical frequency synthesizer design. A slightly underdamped loop can be beneficial to keep the optimal settling time when the process variation is significant. When using an underdamped loop, the overshoot has to be kept within the dynamic range of the charge-pump and the tuning range of the VCO. If the dynamic range of the charge-pump is severely limited, as in a low-voltage design, an overdamped loop can be a better choice to minimize the overshoot. However, the loop bandwidth has to be increased to compensate for the degraded settling time due to overdamping.

4.3. *Settling Time*

Settling time is another important performance metric that is directly related to the loop transfer function. Settling time determines how fast the frequency synthesizer can change the frequency of its output signal.

An analytical solution for the settling time can be obtained from the step response of the closed-loop transfer function, see Eq. (10). Settling time is a function of the natural frequency ($\omega_n = \sqrt{K_D K_o}$) and the damping factor ($\zeta = \omega_n/(2\omega_z)$). It can be shown that

$$t_s \simeq \begin{cases} \dfrac{1}{\zeta\omega_n} \ln \dfrac{\Delta f}{\alpha f_o \sqrt{1 - \zeta^2}} & \text{if } \zeta < 1 \text{ (under)} \\[2ex] \dfrac{1}{\zeta\omega_n} \ln \dfrac{\Delta f}{\alpha f_o} & \text{if } \zeta = 1 \text{ (critical)} \\[2ex] \dfrac{1}{(\zeta - \sqrt{\zeta^2 - 1})\omega_n} \ln \dfrac{\Delta f(\sqrt{\zeta^2 - 1} + \zeta)}{2\alpha f_o \sqrt{\zeta^2 - 1}} & \text{if } \zeta > 1 \text{ (over)} \end{cases} \tag{17}$$

where f_o is the frequency from which the synthesizer starts the transition, Δf is the amount of frequency jump, and α is the settling accuracy. As the loop bandwidth ω_c increases, the settling time gets shorter if the damping ratio is fixed. The effect of the damping ratio on settling time is shown in Fig. 11. It is a plot of Eq. (17) with ω_c fixed but not ω_n, which is more realistic in the sense of design procedure. In this condition, the settling time is fastest when the loop is critically damped, and further underdamping does not improve the settling time. Note that the analytic solution in Eq. (17) is only an approximated result for the second-order closed-loop transfer function, but not for the third-order one. Since Eq. (10) does not take into account the effect of the additional pole, the actual settling time is longer than the analytic solution may suggest.

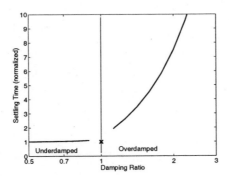

Fig. 11. Settling time vs. damping factor for a second-order PLL

The transient step responses of the third-order transfer functions are shown in Fig. 12(a). The closed-loop transfer functions of the examples are:

$$H_{under}(s) = \frac{1+2s}{1+2s+2s^2+s^3} \tag{18}$$

$$H_{critical}(s) = \frac{1+4s}{1+4s+4s^2+s^3} \tag{19}$$

$$H_{over}(s) = \frac{1+8s}{1+8s+8s^2+s^3} \tag{20}$$

Underdamping is not desirable since it increases overshoot in transient response while not improving settling time performance considerably. The overshoot should be limited within the dynamic range of the charge-pump output, otherwise the settling time performance will be degraded.

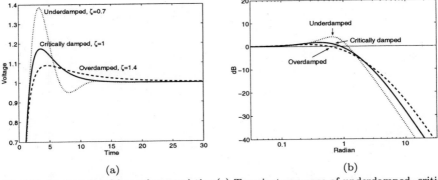

(a)   (b)

Fig. 12. Third-order closed loop characteristics (a) Transient response of underdamped, critically damped and overdamped system (b) Magnitude plot of the closed-loop transfer function shows peaking in underdamped system.

The overshoot in transient response also translates into gain peaking in the frequency domain. Fig. 12(b) shows that an underdamped system has excessive gain peaking due to the stabilizing zero. The gain peaking amplifies the phase noise of the reference signal at the output of the frequency synthesizer. It is recommended to use an overdamped system if the close-in phase noise of the reference signal has considerable effect on the receiver performance.

Finally, the trade-offs of design choices are summarized in Table 3. Loop bandwidth and damping ratios have to be determined carefully, depending on the requirement of the target application, since they improve some aspects of the performance, and deteriorate others at the same time.[c]

Table 3. Summary of PLL design trade-offs

	Loop bandwidth	Damping
Faster settling	wide	under
Better stability	narrow	over
Lower phase noise	wide	N/A
Better spur rejection	narrow	N/A
Low jitter peaking	N/A	over
Low overshoot	N/A	over
Smaller capacitor size	wide	N/A

4.4. *WLAN 802.11b Design Example*

In this section, an example of the design procedure of the frequency synthesizer compliant for Wireless LAN 802.11b standard is presented. The procedure details the considerations for stability and settling time of loop filter design. The same procedure can be applied to different communication standards with minimal modifications.

(i) The first step is to determine the reference frequency f_{REF}. For 802.11b standard, the output frequency must cover the range from 2412 MHz to 2472 MHz in steps of 5 MHz. If the quadrature outputs are to be generated by a divide-by-two circuit, the VCO output frequency has to be twice the requirement. Now the system must cover the range from 4824 MHz to 4944 MHz in steps of 10 MHz. Since $GCD(4824, 10) = 2$, the maximum f_{REF} possible is 2 MHz.

---

[c]For instance, in a frequency synthesizer design, the loop bandwidth is $f_c = 830$kHz and the damping factor is $\zeta = 0.75$, while the reference frequency is $f_{REF} = 11.75$MHz.[16] Since the loop bandwidth is close to the maximum of the Gardner's limit and the damping is underdamped, the PLL shows a fast settling time performance of 40 μs. However, stability of the system is easily disturbed during the measurement and the transient response waveform shows a large overshoot and ringing. In another design example, the loop bandwidth is $f_c = 45$kHz and the damping factor is $\zeta = 1$, while the reference frequency is $f_{REF} = 26.6$MHz. Relatively low loop bandwidth leads to a slow settling time of 250 μs.[14]

(ii) From the Gardner's stability limit, the loop bandwidth ω_c has to be well below ω_{REF}. Considering that the settling time requirement is relatively relaxed, it is beneficial to make the loop bandwidth very narrow to reduce reference spur. Let $\omega_c = 2\pi \times 30\text{kHz}$, then the loop bandwidth is 66 times below f_{REF}.

(iii) For optimal settling time performance, make the loop critically damped. The damping ratio is $\zeta = 1$.

(iv) From Eq. (10) and Eq. (12), the natural frequency is $\omega_n = \omega_c/(2\zeta) = 2\pi \times 15\text{kHz}$.

(v) Now that ω_n and ζ are determined, the settling time can be estimated from the closed form Eq. (17). Using $f_o = 4824$, $\Delta f = 120$ and $\alpha = 25 \times 10^{-6}$, the estimated settling is $t_s = 73$ μs. It is faster than the required 224 μs by a good margin.

(vi) From the loop bandwidth and the damping factor, the location of the stabilizing zero can be determined as $\omega_z = \omega_c/(4\zeta^2) = 2\pi \times 7.5\text{kHz}$

(vii) For a good reference spur rejection performance, it is best to place the additional pole as close to the crossover frequency as possible without degrading phase margin. The optimal location of the additional pole is $\omega_p = \omega_c \times (4\zeta^2) = 2\pi \times 120\text{kHz}$

(viii) Assuming the VCO gain $K_o = 2\pi \times 300\text{MHz/V}$, the PFD-CP gain is $K_D = \omega_n^2/K_o = 4.7\text{V/rad}$.

(ix) Assuming the charge-pump current $I = 30\mu\text{A}$, the rest of the circuit elements can be calculated as follows:
$C_1 = I/(2\pi K_D N) = 420\text{pF}$
$R_1 = 1/(\omega_z C_1) = 50.5\text{k}\Omega$
$C_2 = 1/(\omega_p R) = 26.3\text{pF}$

5. Recent Progress in Frequency Synthesizer Design Techniques

Even though frequency synthesizer theory is very mature, there is still a large research effort aimed to improve performance and optimize implementations for new technologies and emerging standards. One of the main drivers for research in frequency synthesizers has been the need to generate increasingly higher frequencies while decreasing power consumption. This section presents a brief review of recent advances in frequency synthesizer design.

5.1. *Novel Architectures*

The frequency synthesizer architecture is generally based on a phase-locked loop. Dual loop architectures have been presented trying to alleviate the trade-off between loop bandwidth and frequency steps in integer synthesizers.[17,18] An area and power consumption penalty is paid for the relaxed trade-off. A nested architecture is proposed to obtain a wide-band PLL while maintaining fine frequency resolution and spurs rejection.[19] A stabilization technique introduces a zero in the open-loop transfer function through the use of a discrete-time delay cell and relaxes the trade-off between the settling speed and the magnitude of output sidebands.[20]

5.2. *Linearization Techniques*

In an effort to reduce spurious tones, a new topology uses charge-pump averaging and reduces the magnitude of the fractional spurs to levels below the noise floor.[21] A DAC controlled a phase noise cancellation and charge pump linearization technique is introduced.[22] Another option for charge pump linearization is to add a replica charge pump and a bias controller to compensate the current mismatch in the charge pump.[23] This technique allowed a reduction of 8.6 dB of the spurious tones.

5.3. *Digital Phase-Locked Loop*

With the improvement of digital CMOS processes, there has been an increased interest in all-digital RF frequency synthesizers.[24,25,26] One of the main advantages of all-digital frequency synthesizers is the elimination of the PFD - charge pump non linearity, the easy integration in modern technologies and a reduced dependence on process variations. A digital PLL with a DAC to control the VCO voltage and a digital phase-frequency detector (DPFD) accompanied by an adaptive loop control helps to obtain fast acquisition.[24] This frequency synthesizer is mainly oriented to clock generation.

5.4. *Fast Settling Techniques*

Fast settling techniques try to relax the trade-off between settling time and loop bandwidth by providing additional means to speed the frequency switching process. A switchable-capacitor array that tunes the output frequency, and a dual loop filter operating in the capacitance domain are proposed.[27] A settling time smaller than 100 μs is obtained. A locking time as short as 30 μs is reported, which uses a discrete-time loop filter with a stabilization zero created in the discrete-time.[28] A different technique is reported where 64 identical charge pumps are enabled and the loop resistor is reduced by 8×, effectively increasing the loop bandwidth by 8× only during the switching of the synthesizer. A settling time of 10 μs is reported.[29]

5.5. *VCO*

RF oscillator design is challenging due to the uncertainty in the modelling of its passive devices. Hence, it is the building block that has received more attention in the last few years. A phase noise of −139 dBc/Hz at 3 MHz offset is reported using a low inductor quality factor (Q) of 6 for an oscillation frequency of 1.8 GHz in a noise shifting differential Colpitts VCO.[30] −139 dBc/Hz at 3 MHz offset at 1.7 GHz is achieved by adding a voltage regulator to the VCO and thus reducing its sensitivity to the supply noise.[31] A 36 GHz VCO,[32] 60 GHz and 100 GHz VCOs in 90 nm technology,[33] and a 63 GHz VCO in standard 0.25 μm CMOS technology[34] are reported. Circular-geometry oscillators based in slab inductors,[35] and a circular standing wave oscillatorn[36] are presented. A stable fine-tuning loop is combined with an unstable coarse-tuning loop in parallel, and as a result, a stable PLL with a relatively wide tuning range of 600 MHz for a 4.3 GHz oscillator is obtained,[37] and a 20 GHz VCO with 25% tuning range achieved through the small parasitic capacitance of a negative-resistance cell.[38] A single loop horseshoe inductor with a quality factor larger than 20 and an accumulation MOS varactor with Cmax/Cmin ratio of 6 provide a 58.7% tuning range between 3 and 5.6 GHz.[39] Finally, the first digitally controlled oscillator (DCO) incorporating dithering to increase the frequency resolution of the DCO is introduced.[40]

5.6. *Quadrature Generation*

Quadrature generation is an important part of the signal processing in an RF front-end. Most of the modern communication standards use phase or frequency modulation schemes, which require quadrature mixing to extract the information contained in both sides of the spectra.[6]

The most widely used technique involves the use of passive polyphase networks conformed of integrated resistors and capacitors. To improve the accuracy of the 90° phase shift, the order of the phase shift network has to be increased to spread the absolute value of the passive components. Phase errors as low as 3° can be obtained due to process variations of the passive elements.[41,42,43] A drawback of this technique is that the higher the order of the polyphase network, the larger the insertion loss of the LO signal – 3 dB of attenuation per stage. Another common technique for quadrature signal generation is the use of a VCO signal generated at twice the desired LO frequency. This technique provides a broadband range of quadrature outputs, but increases the power consumption by 20 to 30% due to higher operating frequencies. The accuracy of the phase generation is limited by the matching of the flip-flops in the frequency divider and the duty cycle error of the VCO output.[44]

Calibration techniques are also found in the literature; they measure the phase imbalance of the quadrature outputs and compensate it. A delay locked loop (DLL) is used to adjust the phase error in a quadrature generator.[45] A phase detector controls the current in the phase shifter and adjusts the phase different between two split paths. The duty cycle of the clock signal is changed to compensate for the phase imbalance at the output of the divide-by-two circuit by adding a DC level component to the flip-flop clock.[46,47] A self-calibration loop tunes each branch of the phase shifter sequentially to average the phase error generated due to mismatches in the passive components.[48]

5.7. *Prescaler*

Being one of the most power hungry blocks in the synthesizer, along with the VCO, a lot of effort has been placed into reducing its power consumption. Dynamic-logic frequency dividers based on true-single-phase-clock (TSPC) latches have shown a low power and high speed operation.[49] Exploiting dynamic loading, an 1 V 2.5 mW divide-by-two flip-flop operating up to 5.2 GHz in 0.35 μm CMOS technology is achieved.[50] A very low power divider is based on a quasi-differential locking divider operating up to 4.3 GHz while consuming 44 μW from a 0.7 V power supply in a 0.35 μm CMOS process.[51] Another approach to improve power consumption is to use the injection-locked oscillator as a frequency divider.[52] It is shown that the injection-locked frequency divider can provide a high speed divide-by-two circuit with substantially lower power consumption than its digital counterparts.

As can be seen from the previous list of highlighted papers, there are open problems in almost every major building block of the frequency synthesizer. In particular, new architectures that allow to relax the bandwidth and settling time trade-offs, and optimization of VCO performance, along with power efficient frequency dividers, are areas for research focus.

6. Conclusion

A description of frequency synthesizers that emphasizes the key design parameters and specifications for their use in wireless applications has been presented. The mapping between the communication standard into particular specifications has been highlighted for parameters such as phase noise, settling time, and spurious rejection. A discussion on stability limits has been presented to establish the limits on the ratio of loop bandwidth with respect to the reference frequency and the relative location of the poles, zero and crossover frequency. The main design trade-offs between noise, bandwidth and stability have been described, as well as the implications on settling time and stability of the relative location of the pole and zero on the transfer function. A brief survey of the latest advances on the design of frequency synthesizers helps to identify the areas where most of the design effort needs to be put to improve the performance of the circuit. As advances in technology allow for faster and smaller transistors, the trade-offs in the design of frequency synthesizers need to be studied and exploited in the never ending search for a compact and low power transceiver implementation.

Acknowledgment

The authors would like to thank Dr. Gabriele Manganaro, at National Semiconductor; Dr. Keliu Shu, at Texas Instrument; and Alberto Valdes-Garcia, at Texas A&M University, for their invaluable contributions in the form of technical discussions.

References

1. *Wireless LAN Medium Access Control (MAC) and Physical Layer (PHY) Specifications: High-speed Physical Layer in the 5 GHz Band*, IEEE Std. 802.11a, 1999.
2. W. F. Egan, *Frequency Synthesis by Phase Lock.* New York: John Wiley, 1981.

3. *Wireless LAN Medium Access Control (MAC) and Physical Layer (PHY) Specifications: Higher-Speed Physical Layer Extension in the 2.4 GHz Band*, IEEE Std. 802.11b, 1999.

4. *Wireless LAN Medium Access Control (MAC) and Physical Layer (PHY) Specifications: Amendment 4: Further Higher Data Rate Extension in the 2.4 GHz Band*, IEEE Std. 802.11g, 2003.

5. *Specification of the Bluetooth System*, http://www.bluetooth.com Std. v1.0 B, 1999.

6. B. Razavi, *RF Microelectronics*. Upper Saddle River, NJ: Prentice Hall PTR, 1998.

7. T. A. D. Riley, M. A. Copeland, and T. A. Kwasniewsky, "Sigma-Delta Modulation in Fractional-N Frequency Synthesis," *IEEE J. Solid-State Circuits*, vol. 28, pp. 553–559, May 1993.

8. G. Chang, A. Rofougaran, M. Ku, A. A. Abidi, and H. Samueli, "A Low-power CMOS Digitally Synthesized 0-13MHz Agile Sinewave Generator," in *IEEE International Solid-State Circuits Conference, Dig. Tech. Papers*, 1994, pp. 32–33.

9. H. T. Nicholas, III and H. Samueli, "A 150-MHz Direct Digital Frequency Synthesizer in 1.25-μm CMOS with −90-dBc Spurious Performance," *IEEE J. Solid-State Circuits*, vol. 26, pp. 1959–1969, Dec. 1991.

10. A. Yamagishi, M. Ishikawa, T. Tsukahara, and S. Date, "A 2-V, 2-GHz Low-Power Direct Digital Frequency Synthesizer Chip-Set for Wireless Communication," *IEEE J. Solid-State Circuits*, vol. 33, pp. 210–217, Feb. 1998.

11. U. L. Rohde, *Digital PLL Frequency Synthesizers: Theory and Design*. Eaglewood Cliffs, NJ: Prentice Hall, 1982.

12. J. A. Crawford, *Frequency Synthesizer Design Handbook*. Norwood, MA: Artech House, 1994.

13. F. M. Gardner, "Charge-Pump Phase-Lock Loops," in *IEEE Transactions on Communications*, vol. COM-28, Nov. 1980, pp. 1849–1858.

14. J. Craninckx and M. S. J. Steyaert, "A Fully Integrated CMOS DCS-1800 Frequency Synthesizer," *IEEE J. Solid-State Circuits*, vol. 33, pp. 2054–2065, Dec. 1998.

15. K. Shu, E. Sánchez-Sinencio, J. Silva-Martinez, and S. H. K. Embabi, "A 2.4-GHz Monolithic Fractional-N Frequency Synthesizer with Robust Phase-switching Prescaler and Loop Capacitance Multiplier," *IEEE J. Solid-State Circuits*, vol. 38, pp. 866–874, June 2003.

16. C. Lam and B. Razavi, "A 2.6-GHz/5.2-GHz Frequency Synthesizer in 0.4-μm CMOS Technology," *IEEE J. Solid-State Circuits*, vol. 35, pp. 788–794, May 2000.

17. T. Kan and H. C. Luong, "A 2-V 1.8-GHz Fully Integrated CMOS Frequency Synthesizer for DCS-1800 Wireless Systems," in *Symposium on VLSI Circuits*, 2000, pp. 234–237.

18. W. Yan and H. C. Luong, "A 2-V 900-MHz Monolithic CMOS Dual-Loop Frequency Synthesizer for GSM Wireless Receivers," *IEEE J. Solid-State Circuits*, vol. 36, pp. 204–216, Feb. 2001.

19. A. N. Hafez and M. I. Elmasry, "A Fully-Integrated Low Phase-Noise Nested-Loop PLL for Frequency Synthesis," in *Custom Integrated Circuits Conference*, 2000, pp. 589–592.

20. T. C. Lee and B. Razavi, "A Stabilization Technique for Phase-Locked Frequency Synthesizers," in *VLSI Symposium*, 2001, pp. 39–42.

21. Y. Koo, H. Huh, Y. Cho, J. Lee, J. Park, K. Lee, D. Jeong, and W. Kim, "A Fully Integrated CMOS Frequency Synthesizer with Charge-Averaging Charge Pump and Dual-Path Loop Filter for PCS and Cellular CDMA Wireless Systems," *IEEE J. Solid-State Circuits*, vol. 37, pp. 536–542, May 2002.

22. S. Pamarti, L. Jansson, and I. Galton, "A Wideband 2.4-GHz Delta-Sigma Fractional-N PLL With 1-Mb/s In-Loop Modulation," *IEEE J. Solid-State Circuits*, vol. 39, pp. 49–62, Jan. 2004.

23. H. Huh, Y. Koo, Y. Cho, J. Lee, J. Park, K. Lee, D. Jeong, and W. Kim, "A CMOS Dual-Band Fractional-N Synthesizer with Reference Doubler and Compensated Charge Pump," in *IEEE International Solid-State Circuits Conference, Dig. Tech. Papers*, 2004, pp. 100–101.

24. I. Hwang, S. Lee, S. Lee, and S. Kim, "A Digitally Controlled Phase-Locked Loop with Fast Locking Scheme for Clock Synthesis Application," in *IEEE International Solid-State Circuits Conference, Dig. Tech. Papers*, 2000, pp. 168–169.

25. T. Y. Hsu, T. R. Hsu, C. C. Wang, Y. C. Liu, and C. Y. Lee, "Design of a Wide-Band Frequency Synthesizer Based on TDC and DVC Techniques," *IEEE J. Solid-State Circuits*, vol. 37, pp. 1244–1255, Oct. 2002.

26. R. B. Staszewski, C. Hung, K. Maggio, J. Wallberg, D. Lelpold, and P. T. Balsara, "All-Digital Phase-Domain TX Frequency Synthesizer for Bluetooth Radios in 0.13μm CMOS," in *IEEE International Solid-State Circuits Conference, Dig. Tech. Papers*, 2004, pp. 272–273.

27. C. Lo and H. C. Luong, "A 1.5-V 900-MHz Monolithic CMOS Fast-Switching Frequency Synthesizer for Wireless Applications," *IEEE J. Solid-State Circuits*, vol. 37, pp. 459–470, Apr. 2002.

28. B. Zhang, P. E. Allen, and J. M. Huard, "A Fast Switching PLL Frequency Synthesizer with an On-Chip Passive Discrete-Time Loop Filter In 0.25μm CMOS," *IEEE J. Solid-State Circuits*, vol. 38, pp. 855–865, June 2003.

29. M. Keaveney, P. Walsh, M. Tuthill, C. Lyden, and B. Hunt, "A 10μs Fast Switching PLL Synthesizer for GSM/EDGE Base Station," in *IEEE International Solid-State Circuits Conference, Dig. Tech. Papers*, 2004, pp. 272–273.

30. R. Aparicio and A. Hajimiri, "A Noise-Shifting Differential Colpitts VCO," *IEEE J. Solid-State Circuits*, vol. 37, pp. 1728–1736, Dec. 2002.

31. Y. Wu and V. Aparin, "A Monolithic Low Phase 1.7GHz CMOS VCO for Zero-IF Cellular CDMA Receivers," in *IEEE International Solid-State Circuits Conference, Dig. Tech. Papers*, 2004, pp. 396–397.

32. H. Veenstra and E. van der Heijden, "A 35.2 37.6GHz VCO in a 70/100GHz ft/fmax SiGe Technology," in *IEEE International Solid-State Circuits Conference, Dig. Tech. Papers*, 2004, pp. 394–395.

33. L. M. Franca-Neto, P. E. Bishop, and B. A. Bloechel, "64GHz and 100GHz VCOs in 90nm CMOS Using Optimum Pumping Method," in *IEEE International Solid-State Circuits Conference, Dig. Tech. Papers*, 2004, pp. 444–445.

34. R. C. Liu, H. Y. Chang, C. H. Wang, and H. Wang, "A 63 GHz VCO Using a Standard 0.25 μm CMOS Process," in *IEEE International Solid-State Circuits Conference, Dig. Tech. Papers*, 2004, pp. 446–447.

35. R. Aparicio and A. Hajimiri, "Circular-Geometry Oscillators," in *IEEE International Solid-State Circuits Conference, Dig. Tech. Papers*, 2004, pp. 378–379.

36. D. Ham and W. Andress, "A Circular Standing Wave Oscillator," in *IEEE International Solid-State Circuits Conference, Dig. Tech. Papers*, 2004, pp. 380–381.

37. G. F. F. Herzel and H. Gustat, "An Integrated CMOS RF Synthesizer for 802.11a Wireless LAN," *IEEE J. Solid-State Circuits*, vol. 38, pp. 1767–1770, Oct. 2003.

38. B. Jung and R. Harjani, "A 20 GH VCO with 5 GHz Tuning Range in 0.25 μm SiGe BiCMOS," in *IEEE International Solid-State Circuits Conference, Dig. Tech. Papers*, 2004, pp. 178–179.

39. N. H. W. Fong, J. Plouchart, N. Zamdmer, D. Liu, L. Wagner, C. Plett, and N. G. Tarr, "Design of Wide-Band CMOS VCO for Multiband Wireless LAN Applications," *IEEE J. Solid-State Circuits*, vol. 38, pp. 1333–1342, Aug. 2003.

40. R. B. Staszewski, D. Lelpold, K. Muhammad, and P. T. Balsara, "Digitally Con-

trolled Oscillator (DCO) Based Architecture for RF Frequency Synthesis in a Deep-Submicrometer CMOS Process," *IEEE Trans. Circuits Syst. II*, vol. 50, pp. 815–828, Nov. 2003.

41. F. Behbahani, Y. Kishigami, J. Leete, and A. A. Abidi, "CMOS Mixers and Polyphase Filters for Large Image Rejection," *IEEE J. Solid-State Circuits*, vol. 36, pp. 873–887, June 2001.

42. J. Crols and M. S. J. Steyaert, "A Single-Chip 900 MHz CMOS Receiver Front-End with a High Performance Low-IF Topology," *IEEE J. Solid-State Circuits*, vol. 30, pp. 1483–1492, Dec. 1995.

43. F. O. Eynde, J. Schmit, V. Charlier, R. Alexandre, C. Sturman, K. Coffin, B. Mollekens, J. Craninckx, S. Terrijn, A. Monterastelli, S. Beerens, P. Goetschalckx, M. Ingels, D. Joos, S. Guncer, and A. Pootioglu, "Fully Integrated Chip SOC for Bluetooth," in *IEEE International Solid-State Circuits Conference, Dig. Tech. Papers*, 2001, pp. 196–197.

44. J. P. Maligeorgos and J. R. Long, "A Low-Voltage 5.1-5.8-GHz Image-Reject Receiver with Wide Dynamic Range," *IEEE J. Solid-State Circuits*, vol. 35, pp. 1917–1926, Dec. 2000.

45. D. Lovelace and J. Durec, "A Self Calibrating Quadrature Generator with Wide Frequency Range," in *IEEE Radio Frequency Integrated Circuits Symposium*, 1997, pp. 147–151.

46. F. Behbahani, "An adjustable Bipolar Quadrature LO Generator with an Improved Divide-by-2 Stage," in *Bipolar/BiCMOS Circuits and Technology Meeting*, Oct. 1996, pp. 157–160.

47. S. Navid, F. Behbahani, A. Fotowat, A. Hajimiri, R. Gaethke, and M. Delurio, "Level-Locked Loop: A Technique for Broadband Quadrature Signal Generation," in *IEEE Custom Integrated Circuits Conference*, May 1997, pp. 411–414.

48. S. H. Wang, J. Gil, I. Kwon, H. K. Ahn, H. Shin, and B. Kim, "A 5-GHz Band I/Q Clock Generator Using a Self-Calibration Technique," in *European Solid-State Circuits Conference*, 2002, pp. 807–810.

49. S. Pellerano, S. Levantino, C. Samori, and A. Lacaita, "A 13.5 mW 5 GHz Frequency Synthesizer with Dynamic Logic Frequency Divider," *IEEE J. Solid-State Circuits*, vol. 39, pp. 378–383, Feb. 2004.

50. J. M. C. Wong, V. S. L. Cheung, and H. Luong, "A 1 V 2.5 mW 5.2 GHz Frequency Divider in a 0.35 μm CMOS Process," *IEEE J. Solid-State Circuits*, vol. 38, pp. 1643–1648, Oct. 2003.

51. K. Yamamoto and M. Fujishima, "4.3 GHz 44 μW CMOS Frequency Divider," in *IEEE International Solid-State Circuits Conference, Dig. Tech. Papers*, 2004, pp. 104–105.

52. S. Verma, H. R. Rategh, and T. H. Lee, "A Unified Model for Injection-Locked Frequency Dividers," *IEEE J. Solid-State Circuits*, vol. 38, pp. 1015–1027, June 2003.

International Journal of High Speed Electronics and Systems
Vol. 15, No. 2 (2005) 377–428
© World Scientific Publishing Company

RECENT ADVANCES AND DESIGN TRENDS IN CMOS RADIO FREQUENCY INTEGRATED CIRCUITS

DAVID J. ALLSTOT, SANKARAN ANIRUDDHAN, MIN CHU,
JEYANANDH PARAMESH, AND SUDIP SHEKHAR[1]

Department of Electrical Engineering, University of Washington
Seattle, WA 98195-2500, United States of America

Several state-of-the-art wireless receiver architectures are presented including the traditional super-heterodyne, the image-reject heterodyne, the direct-conversion, and the very-low intermediate frequency (VLIF). The case studies are followed by a detailed view of receiver building blocks: low-noise amplifiers (LNA), mixers, and voltage-controlled oscillators (VCO). Two popular topologies currently exist for LNAs: the common-gate configuration, which offers low power consumption with superior stability, robustness and linearity performance, and its common-source counterpart, which provides comparatively higher gain and lower noise figure. Aside from the traditional passive and active Gilbert mixers, the even-harmonic and masking-quadrature mixers are developed to combat second-order non-linearity and improve image-rejection, respectively. For quadrature carrier generation, the degeneration-injected QVCO is superior to the cascode-injected QVCO both in terms of phase noise and tuning range. The Colpitts QVCO is attractive as a low-noise alternative as it does not disturb the output voltage as much as its traditional LC counterpart and thus offers lower phase noise.

Keywords: Super heterodyne transceiver; image-reject transceiver; direct-conversion transceiver; low-IF transceiver; low-noise amplifier; RF mixer; voltage-controlled oscillator.

1. Introduction

The wireless industry is experiencing explosive growth with applications like GSM cellular telephony, Bluetooth, and wireless LAN becoming commodities of everyday life. One major factor contributing to this growth is the emerging presence of CMOS technology in commercial analog and radio frequency integrated circuits (RFIC). Traditionally, technologies such as Gallium Arsenide (GaAs) and Silicon Germanium (SiGe) are chosen over CMOS for fabricating RF circuits because of stringent noise, linearity, gain, and power efficiency requirements in addition to the high speed, low power, and high yield requirements characteristic of digital integrated circuits. The drawbacks of these technologies, however, are their high cost and slow scaling needed to achieve high integration. This was evident in the past where improvements in RF circuit performance have been slower than those of the digital microprocessor industry that is guided by Moore's law. Recently, with the drive towards system-on-chip (SOC)

---

[1] Author emails: allstot@ee.washington.edu, ani@ee.washington.edu, adamchu@ee.washington.edu, paramesh@ee.washington.edu, shekhar@ee.washington.edu

integration, there has been a tremendous interest in implementing RF circuits such as low noise amplifiers (LNA), mixers, power amplifiers (PA) and voltage-controlled oscillators (VCO) in digital CMOS technology. The reasons are quite obvious: if CMOS RF circuits are able to provide similar performance as their GaAs and SiGe counterparts, they may experience thenir own version of Moore's law with regard to integration ease and low fabrication cost. Here, we present some recent advances and design trends in CMOS RF ICs at both the architectural and circuit levels. The paper is divided into six sections. Section II addresses key issues in receiver architectures. Section III explores the commonly used LNA topologies along with their advantages and drawbacks. Various mixer and VCO topologies and their implementations are described in Sections IV and V, respectively, and conclusions are given in Section VI.

2.   Radio Receiver Architectures

In this section, we present an overview of radio receiver architectures in an approximate chronological sequence in terms of their evolution and implementation in highly integrated silicon processes. The narrative starts with the classical super-heterodyne architecture followed by its extension to image-reject architectures before moving on to the direct-conversion architecture; in each case, we discuss the advantages and disadvantages in terms of their performance and amenability to silicon integration. This leads naturally into modern radio architectures, almost all of which utilize direct-conversion or very-low intermediate frequency (VLIF) architectures.

A small fraction of modern consumer receivers is based on the sampling technique wherein the desired signal is directly sampled or sub-sampled either at RF or IF; such signals may either be directly digitized using a bandpass analog-to-digital converter (ADC) or filtered and down-converted using discrete-time analog decimation filters prior to digitization. Such receivers have become popular in applications such as radar, which are not constrained by cost, power and a compelling need for integration. In fully integrated form with applications to portable electronics, they suffer from a number of problems that have not been resolved. As a result, they are typically out-performed by traditional architectures. Nevertheless, sampled-data radios are being actively researched, and in some cases used in commercial products. In spite of the resurgent interest in such architectures, we will not describe them in detail herein owing to the tutorial nature of this paper.

2.1. *The super-heterodyne architecture*

The classical super-heterodyne architecture (Fig. 1(a)) has been the workhorse for over 90% of radios built over the last several decades. Conceived by Edwin Armstrong in 1918 [1], the super-heterodyne features high selectivity and sensitivity. Fig. 1(b) depicts the signal flow through the receive chain. The RF signal received at the antenna is first filtered using the pre-select bandpass filter to isolate the desired frequency band. After amplification by a low-noise amplifier, the signal is then fed into an image-reject filter whose function is twofold: first, it greatly attenuates any unwanted signal at the "image frequency" which would otherwise be down-converted to the target intermediate frequency (LO) by the mixer, and second, it also attenuates the thermal noise that accompanies the input signal at the image frequency. The output of the image-reject filter is then down-converted and the resulting IF signal is filtered using the channel-select filter to isolate the desired channel from close-in interferers. At this point, the signal is ready for further amplification and down-conversion to baseband where it may be demodulated as such or after digitization.

The primary design choice in a super-heterodyne is that of the IF. For an input frequency f_{RF} the image frequency is located at $f_{RF} + 2f_{IF}$ for high-side LO injection and $f_{RF} - 2f_{IF}$ for low-side LO injection; in either case, the separation between the RF signal and its image is $2f_{IF}$. When this separation is large (high IF), the image is greatly attenuated; however, close-in interferers are not suppressed with the image-reject filter, thus necessitating a larger dynamic range (DR) for the mixer. Conversely, in the low IF case where the separation is small, the attenuation of the image is poor while the close-in interferers are somewhat attenuated; this choice may alleviate the DR requirement on the mixer at the expense of greater corruption of the desired signal by its image. Thus, there exists an inherent trade-off between the image-reject and adjacent channel suppression requirements, which dictates that both filters be highly selective. Filters with sharp cut-off characteristics and the requisite DR are impractical in modern IC technologies, and have been traditionally implemented as off-chip passive surface-acoustic wave (SAW) or LC filters (e.g., [2]). Aside from the obvious disadvantages in terms of cost and form-factor, these filters typically present 50Ω loads to the chip. Consequently, circuit blocks that would otherwise drive high-impedance capacitive loads must now be buffered to drive these filters.

2.2. *Image-reject heterodyne architectures*

The trade-off between image-rejection and close-in interferer suppression can be alleviated through the use of image-reject architectures featuring desired signal enhancement and image-signal cancellation; examples include the Hartley and Weaver topologies depicted in Fig. 2(a) and Fig. 2(b), respectively.

A brief mathematical description of the Hartley [3] and Weaver [4] image-reject receivers is provided in Fig. 3. In each case, some of the circuitry may be eliminated depending on the desired sideband. Passive poly-phase filters have recently been re-discovered to implement the Hilbert transformation [5].

The image-reject ratio (IRR) is defined as the ratio of the desired signal gain to the image signal gain and is ideally infinite. IRR is limited in practice by the gain mismatch ε and phase mismatch δ (in radians) between the I and Q paths and is given by:

$$IRR = \frac{1 + (1+\varepsilon)^2 + 2(1+\varepsilon)\cos\delta}{1 - (1+\varepsilon)^2 + 2(1+\varepsilon)\cos\delta} \approx \frac{\varepsilon^2 + \delta^2}{4} \tag{1}$$

For a gain mismatch of 1% and a phase mismatch of 1°, values that are typically achieved on an IC when all process, voltage, and temperature (PVT) variations are taken into account, IRR drops to 39dB. While this is sufficient for some well-controlled applications such as GSM, 60dB or more may be desired in standards which operate in unlicensed bands.[2] Furthermore, phase mismatch becomes increasingly worse at higher frequencies. For example, to achieve a phase mismatch of 1° at 1GHz, the I/Q LO signals can tolerate only 2.8ps of mismatch and this margin dwindles to about 0.7ps at 5GHz. Careful attention to the design of the generation of the quadrature LO signals is therefore essential. These are typically generated by one of three methods:

- a quadrature oscillator directly generates the I/Q phases at the target LO frequency
- a normal single-phase (or differential) oscillator operating at the target LO frequency followed by a polyphase filter generates quadrature phases, or
- a single-phase or differential oscillator operating at *twice* the target LO frequency is followed by a divide-by-2 frequency divider to generate the quadrature phases.

---

[2] The problem of I/Q mismatch has implications beyond limiting the image-rejection effectiveness of a Hartley or Weaver receiver. For instance, in a direct-conversion receiver, the desired signal is its own self-image, and I/Q mismatch can cause distortion of the constellation. This can be problematic in standards that use high-order modulation schemes such as 64-QAM, and is discussed in a later section.

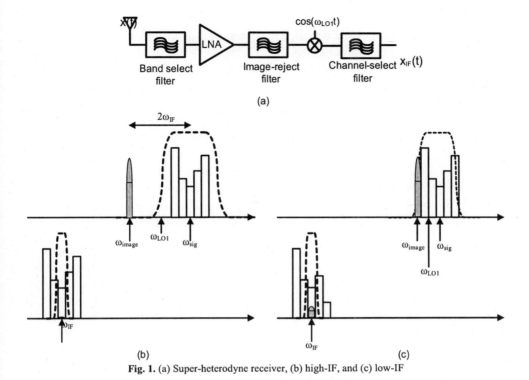

Fig. 1. (a) Super-heterodyne receiver, (b) high-IF, and (c) low-IF

In the image-reject heterodyne receiver, a high IRR is typically achieved by choosing an appropriately high IF [6] such that the band-select filter significantly attenuates the image prior to downconversion. In a later section, we shall describe digital calibration techniques that have been reported in the literature to achieve high IRRs.

The disadvantages of super-heterodyne and image-reject receivers have been greatly magnified by the desire to eliminate all off-chip filters except the band-select filter. This has motivated the resurgence of the direct-conversion architecture in spite of its long history of failure as a high-performance receiver. Before moving to the topic of direct-conversion receivers, we now take a brief detour to present a case-study of a recently reported CMOS heterodyne image-reject receiver.

2.2.1. *Case study*

In order to exemplify some of the issues pertaining to the design of a heterodyne receiver, we briefly describe the Digital European Cordless Telephone (DECT) receiver architecture presented in [7]. This so-called *wide-band double-IF* receiver is a variant of the Weaver image-reject receiver. The input signal occupies the 1881-1897MHz band and is down-converted using a fixed LO to a sliding IF over the range 181-197MHz. The motivation for using a fixed-LO is the requirement of low-phase noise, which could be attained only by use of a wideband PLL; furthermore, having a second tunable PLL at IF to generate a pure second LO for the IF to baseband down-conversion would prove considerably easier. The outputs of the first quadrature mixers are filtered by their first-order RC loads. The IF signal is then down-converted to baseband by a second set of mixers. All mixers use the CMOS variant of the Gilbert cell. The reported chip also has integrated I and Q baseband filters, which are a combination of continuous-time Sallen-Key stages and a switched-capacitor filter followed by 10b ADCs. The overall image rejection obtained was measured to be 70dB, with 35dB provided by the RF band-select filter and the remainder by the image-reject architecture.

2.3. *The direct-conversion architecture*

The direct-conversion receiver (Fig. 3) is conceptually the simplest architecture. The image problem is bypassed because the LO frequency is equal to that of the RF carrier. The advent of integer-N synthesizers followed by the more sophisticated fractional-N frequency synthesizers has made it possible to accurately tune the LO in extremely small steps corresponding to the channel spacing. The down-converted signal includes the desired channel accompanied by a large number of close-in interferers. This information can be further processed in one of two ways: one, channel-selection can be performed by a high-order analog filter prior and the resulting signal digitized with a low-resolution ADC, or alternatively, a low-order anti-aliasing filter may be used in conjunction with a high-resolution ADC so that channel-selection is relegated to the digital domain. Both options open up the possibility of full-integration of the baseband filtering function. This is an attractive proposition because the task of implementing highly selective *monolithic* filters with large DR is considerably easier at baseband than at a high IF. Because the entire receive chain has been integrated, the LNA and the mixer no longer need to drive 50Ω loads which potentially leads to significant power savings.

In spite of these compelling advantages, the direct-conversion receiver has proven the most difficult to build. Not surprisingly, the associated problems can be traced to DC, and as such are either non-existent or of minor importance in a heterodyne receiver.

2.3.1. *LO self-mixing and DC offsets*

The problem of LO self-mixing, depicted in Figure 4, arises because the LO in the direct-conversion receiver is tuned to exactly the RF carrier frequency. The large LO signal may radiate out of the receiver and couple back into the receive antenna either directly or through a nearby reflector. This parasitic leakage is amplified by the LNA and appears at the RF port of the mixer where it mixes with the LO and gets down-converted to DC, thus appearing added to the down-converted signal of interest. This problem is exacerbated if the outward LO leakage is coupled back to the receive antenna through a mobile reflector; in this case, the Doppler shift causes a time-varying DC offset. A second source of parasitic coupling arises due to the finite isolation between the RF and

LO ports, wherein a strong interferer can phase-modulate the LO and self-mix down to the baseband frequencies.

To gain an appreciation for the severity of the problem, consider an application of direct-conversion to the GSM standard. Suppose that the LNA and mixer together have a gain of 30dB and the baseband chain has a maximum gain of 70dB. The desired RF signal can be lower than 3.16μV peak (-100dBm) while the LO port is typically driven by a large signal on the order of 300mV peak (0dBm). When the parasitic LO leakage appearing at the antenna is 60dB lower than the LO, the DC offset at the mixer output is 10mV while the down-converted signal level is about 100μV.

In addition to the DC offset caused by self-mixing, ever-present mismatches in the baseband amplifiers and filters cause an equivalent input-referred DC offset to compete with the down-converted signal. In typical CMOS circuits, this offset can be on the order of a few millivolts while the down-converted baseband signal is usually in the range of a few hundred microvolts. Therefore, in the absence of a DC offset cancellation scheme, the signal-to-noise ratio (SNR) at the detector input can be seriously degraded. Worse yet, the receiver may saturate due to the large gain in the baseband chain multiplying the undesired DC offset voltage.

Another cause for DC offset at the mixer output arises from even-order distortion. For example, suppose that a close-in interferer $v_i = A_1 \cos \omega_1 t$ is present at the baseband amplifier input whose second-order non-linearity is modelled as $v_o = a_1 v_i + a_2 v_i^2$. The output of the amplifier is then $v_o = a_1 A_1 \cos \omega_1 t + 0.5 a_2 A_1^2 (1 + \cos 2\omega_1 t)$, which contains a DC offset dependence on the amplitude of the interferer. Although the second-order non-linearity of a fully differential circuit is theoretically zero, device and layout mismatches set a practical finite limit.

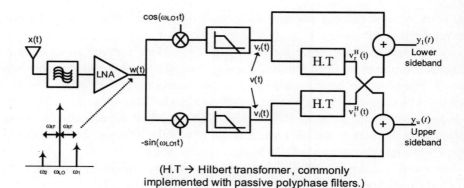

(H.T → Hilbert transformer, commonly
implemented with passive polyphase filters.)

LNA Output : $w(t) = A_1 \cos\omega_1 t + A_2 \cos\omega_2 t$

Filtered IF : $v(t) = v_r(t) + jv_r(t) = \dfrac{A_1}{2}e^{j\omega_{IF}t} + \dfrac{A_2}{2}e^{-j\omega_{IF}t}$

Hilbert Transforms :
$\begin{cases} v_r(t) = \left(\dfrac{A_1}{2} + \dfrac{A_2}{2}\right)\cos\omega_{IF}t \Rightarrow v_r^H(t) = \left(\dfrac{A_1}{2} + \dfrac{A_2}{2}\right)\sin\omega_{IF}t \\[2mm] v_i(t) = \left(\dfrac{A_1}{2} - \dfrac{A_2}{2}\right)\sin\omega_{IF}t \Rightarrow v_i^H(t) = -\left(\dfrac{A_1}{2} - \dfrac{A_2}{2}\right)\cos\omega_{IF}t \end{cases}$

Lower sideband output : $y_1(t) = v_r(t) + v_i^H(t) = A_2 \cos\omega_{IF}t$

Upper sideband output : $y_u(t) = v_i(t) + v_r^H(t) = A_1 \sin\omega_{IF}t$

(a)

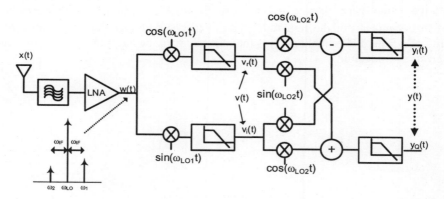

LNA Output : $w(t) = A_1 \cos\omega_1 t + A_2 \cos\omega_2 t$

Filtered first IF : $v(t) = v_r(t) + jv_r(t) = \dfrac{A_1}{2}e^{j\omega_{IF}t} + \dfrac{A_2}{2}e^{-j\omega_{IF}t}$

Filtered second IF : $y(t) = y_1(t) + jy_Q(t) = A_1 \cos\omega_{IF}t + A_1 \sin\omega_{IF}t$

This extracts the upper-sideband. When $\omega_{IF2} = 0$, both $y_1(t)$ and $y_Q(t)$ are required.

For $\omega_{IF2} \neq 0$, one of the two is sufficient.

To extract the lower side-band, the signs in the top and bottom summers
following the second set of mixers must be reversed.

(b)

Fig. 2. (a) Hartley, and (b) Weaver receiver

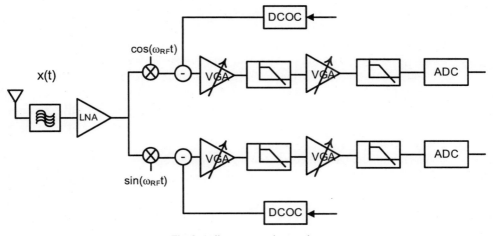

Fig. 3. A direct-conversion receiver

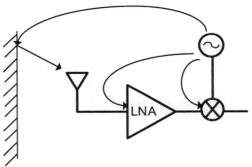

Fig. 4. LO self-mixing due to re-radiation

Several solutions to solving the DC offset problem have been proposed:

- A DC-free modulation scheme may be used. A typical example is the frequency-shift keying (FSK) technique used in pagers. In such cases, AC coupling may be used without significant loss of data. Another example is the orthogonal frequency division multiplexing (OFDM) technique used in the 802.11a standard where the sub-carrier centered at DC is left un-modulated. Since each subcarrier is separated by 312kHz, AC coupling with a cut-off frequency of less than 156kHz may be used.

- A feedback-based DC offset cancellation scheme may be implemented. The baseband chain often consists of cascade of several stages of variable gain amplification and low-pass filtering. Each stage may have its own local DC cancellation loop [7], or a global digital cancellation loop may be employed [8].

- DC offsets caused by LO self-mixing are solved using an intelligent frequency plan. For example, in Conexant's quad-band GSM receiver [7], the high-band

RF signal occupies the 1800/1900 MHz bands while the low-band RF signal occupies the 800/900MHz bands. The VCO is tunable over 1250-1650MHz, and its frequency is divided by 3 to produce a LO over the range 415MHz-550MHz. This LO is used in conjunction with an *even harmonic mixer* (described later) to directly down-convert the low-band RF input while a frequency doubler is used to obtain the LO that down-converts the high-band RF input. Since the LO operates at ½ or ¼ the RF frequency, the mixing of the re-radiated LO does not mix down to DC. The same argument may be made for re-radiation of the main VCO signal.

- To deal with the DC offsets caused by even-order distortion, digitally-controlled calibration schemes that introduce intentional asymmetry in the differential paths have been proposed [7].

2.3.2. *Flicker noise*

Flicker noise is a significant problem in CMOS direct-conversion receivers, especially in applications using modulation schemes whose spectra peak at DC. Bipolar junction transistor (BJT) implementations also suffer from a flicker noise problem, albeit to a much lesser extent. For this reason, GSM direct-conversion receivers have been reported only in BJT technologies [8]-[10]. However, many recently reported 802.11a receivers [11], [12] employ direct-conversion because the sub-carrier centred around DC does not carry data. Where possible, auto-zero and correlated double sampling schemes have been employed to minimize flicker noise [13].

2.3.3. *Even-order non-linearities*

Heterodyne receivers are typically limited by odd-order distortion products. However, direct-conversion receivers additionally suffer from even-order distortion. To understand how even-order distortion affects direct-conversion receivers, consider the non-linear transfer function, $v_o = a_0 v_i + a_1 v_i^2$, with an amplitude modulated interferer $v_i = a(t)cos\omega_i t$ present at its input. The output then contains a baseband component $a^2(t)$ because of demodulation of the AM interferer. The GSM standard specifies a test for just this scenario: the receiver must continue to operate reliably with a -99dBm RF signal when a -31dBm AM blocker at 6MHz offset turns on in the middle of a receive slot. Even-order non-linearities are characteristic of single-ended circuits, and can occur in differential circuits due to component mismatch.

2.3.4. *I/Q mismatch*

We observed previously that I/Q mismatch is somewhat less of an issue in direct-conversion receivers. However, the margin for I/Q mismatch errors falls with higher-order modulation formats and is further exacerbated by higher carrier frequencies. The effect of I/Q mismatches is shown in Fig. 5 for a 64-QAM signal with 5% gain mismatch and 2° phase mismatch, not atypical values. The slicer levels in the demodulator fall along the grids in each axis. Without I/Q mismatch, each transmitted symbol falls in the center of each square. With mismatch errors, the symbols are skewed off each grid center and cause a drop in SNR at the detector input if left uncorrected.

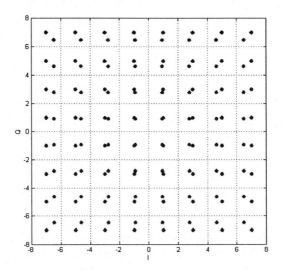

Fig. 5. Effect of I/Q mismatch on a direct-conversion receiver for 64-QAM

2.3.5. *Case study*

The direct-conversion receiver presented in [13] is a state-of-the-art receiver for 5GHz band wireless LANs. The receiver portion of this chip integrates the LNA, quadrature mixers, variable-gain baseband amplifiers and filter, and the synthesizer which operates at ½ the RF frequency to avoid VCO pulling; the target LO frequency is obtained using a frequency-doubler and polyphase filters for quadrature generation. The key features of the receiver are the digital calibration schemes used to correct DC offsets and I/Q mismatches. The output of the companion transmitter in [7] (not described herein) is looped back to measure the I/Q mismatch in the receiver.[3] This measurement is then used to program the coefficients of an adaptive filter which post-distorts the received constellation. Specifically, a constant amplitude, rotating phase calibration signal applied by loop-back appears in the received I-Q plane as an ellipse instead of a circle in the

---

[3] The I/Q mismatch in the transmitter itself is corrected by looping back its output to the ADC input; this is, in effect, a sub-sampling receiver.

presence of calibration. When calibration is performed and post-distortion accomplished, this locus in the I-Q plane closely approximates a circle as shown in Fig. 6.

2.4. *The very-low IF architecture*

The VLIF receiver [14], shown in Fig. 7, seeks to retain the benefits of a direct-conversion receiver while trying to mitigate its more serious shortcomings. The RF signals, along with a number of close-in interferers, are quadrature down-converted to a low IF. All the requisite filtering is performed by fully integrated real or complex lowpass filters. Although it is possible to perform the channel select function in the analog domain, it is typically relegated to the digital domain. Separate ADCs in the I and Q paths digitize the desired signal, which resides at some low IF, along with a number of undesired, partially filtered signals[4]. A complex mixer or Hilbert filter implemented in the digital domain performs image rejection, after which real-coefficient I/Q digital filters isolate the desired channel. Alternatively, the channel selection could be performed by filtering the output of the ADC with a digital filter with complex coefficients.

In spite of a deliberate attempt to combine the advantages of the heterodyne and direct-conversion architectures, the VLIF receiver suffers from several disadvantages. First, the IRR of a typical VLIF receiver is limited to about 40dB by the RF and analog component mismatch in the down-conversion and baseband chain. Furthermore, the low IF often lies within the passband of the RF band-selector which, as such, offers *no* additional attenuation of the image signal. Second, when real I and Q bandpass IF filters are used, the requirements on the ADC are more stringent than in the case of the direct-conversion receiver because each must operate at IF with an attendant power penalty. When a complex IF filter is used, the ADC requirements are the same as in a direct-conversion receiver [14].[5] Finally, the problem of even-order distortion persists in the VLIF receiver, and it may possibly be even worse because of the partial filtering of the blockers. For example, in a GSM receiver using an IF equal to ½ the channel bandwidth, 100kHz, any two blockers separated by one channel could potentially have an even-order intermodulation product at the IF.

2.4.1. *Case study*

A typical example is Motorola's GSM receiver [15] which places the IF at 100kHz, roughly one-half the channel bandwidth. Consequently, the image frequency resides in the adjacent channel (ACI). The key observation behind this choice of IF is that the GSM standard restricts the amplitude of the ACI to be no more than 9dB greater than the

---

[4] Analog filtering and digitization may also be performed with complex (active or passive poly-phase) filters in conjunction with a complex ΣΔ ADC.

[5] The disadvantage of a complex filter is that I/Q mismatch calibration becomes more complicated; when using real filters, the calibration may be accomplished easily using an AGC algorithm.

desired channel. The resulting IRR requirement of about 35dB can be realistically attained by a VLIF image-reject receiver. In [15], the fourth-order baseband filter has a 3dB frequency of about 600kHz; correspondingly, the ADC requires 84dB DR [16].

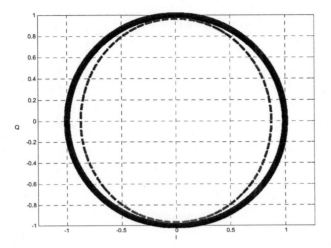

Fig. 6. The effect of I/Q mismatch calibration using a constant-amplitude, rotating-phase calibration signal

3. Low Noise Amplifiers

A low noise amplifier (LNA) is the first active stage of an RF receiver (Fig. 8). The foremost task of an LNA is to amplify the RF signal and suppress noise contributions from subsequent stages. For an *n*-stage cascade, the overall noise factor is given by the Friis equation [18]:

$$F_{Total} = F_1 + \frac{F_2-1}{A_{p1}} + ... + \frac{F_n-1}{\prod_{i=1}^{n-1} A_{pi}} \tag{2}$$

where F_i and A_{pi} are the noise factor and available power gain of the *i*th-stage, respectively (Fig. 9). Thus, high gain (A_{p1}) and low noise factor (F_1) of the first stage (LNA) of an RF receiver are critical for an overall receiver low noise factor (F_{Rx}). Apart from high gain and low noise factor, an LNA needs to present a 50Ω termination at its input to the transmission line from the antenna or the pre-select filter preceding the LNA so the filtering characteristics such as insertion loss and pass band ripple are ensured.

In addition to gain, noise figure, and input matching, other performance parameters critical to an LNA include stability, power consumption, robustness against process, voltage, and temperature (PVT) variations, and linearity, which is typically characterized by the second- or third-order intermodulation product [19].

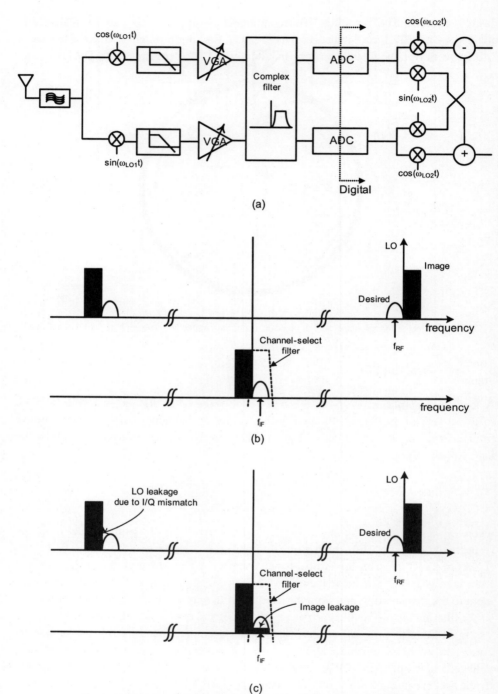

Fig. 7. (a) A VLIF receiver, (b) ideal signal flow, and (c) effects of I/Q mismatch

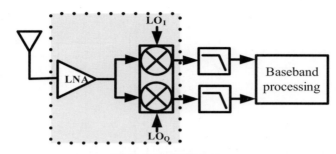

Fig. 8. A simple RF receiver

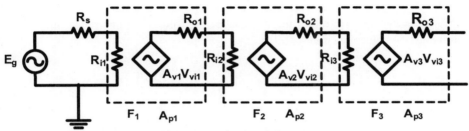

Fig. 9. An *n*-stage cascaded system

The two most widely adopted LNA topologies are inductively degenerated common-source and common-gate amplifiers. While the common-source topology provides higher transconductance and superior noise figure (NF), its common-gate counterpart offers superior reverse isolation (no Miller effect), more robust operation, and lower power consumption. In the following section, the two topologies are compared in detail using the aforementioned performance parameters [20].

3.1. *Common-gate LNA*

3.1.1. *Input matching*

Figure 10 shows a common-gate LNA in which the input impedance, $1/g_m$, is set to 50Ω to obtain the desired matching. The inductor L_s tunes out the parasitic capacitances C_{gs} and C_{sb} at the source node to give purely resistive impedance matching at the operating frequency. This input matching is fairly broadband in nature, suggesting the advantages of using common-gate topologies for wideband applications.

3.1.2. *Noise factor*

The noise factor of a common-gate LNA, under the input matching condition of $1/g_m = R_s$, and ignoring the small contributions from the induced gate noise, is given by:

$$F = 1 + \frac{\overline{i_d^2}}{\overline{i_{Rs}^2}} = 1 + \frac{\gamma}{\alpha} \tag{3}$$

where

$$\overline{i_d^2} = 4kT\gamma \, g_{d0} \Delta f$$
$$\overline{i_{Rs}^2} = 4kT\Delta f / R_s$$
$$\alpha = g_m / g_{d0}$$

Here, α and γ are bias-dependent parameters [21]. The condition of input matching sets a lower bound on the noise factor of the common-gate LNA, and leads to inferior noise performance. However, the noise factor is approximately constant with respect to the operating frequency, suggesting the advantages of using a common-gate topology for high frequency implementations.

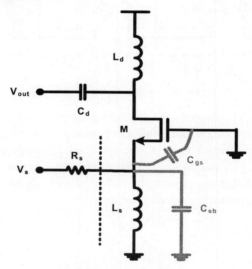

Fig. 10. Common-gate LNA

3.1.3. *Gain*

The effective input transconductance of the common-gate LNA under perfect input matching conditions is:

$$G_{m,eff} = \frac{1}{2} g_m = \frac{1}{2R_s} \tag{4}$$

3.1.4. *Robustness*

Inductor L_s and the parasitic capacitances C_{gs} and C_{sb} at the source node of the common-gate LNA form a parallel resonant RLC tank network with resonant frequency, ω_0. The quality factor of this tank is:

$$Q = \frac{\omega_0 C_{gs} R_s}{2} < 1 \tag{5}$$

Since a lower quality factor of the matching network leads to a lower sensitivity of Z_{in} to parasitic components [22], the common-gate LNA is highly robust against typical PVT variations. Moreover, any parasitic capacitances as well as the pad capacitances are conveniently absorbed into the LC tank.

3.1.5. *Reverse isolation and stability*

The stability factor for an amplifier [23] is given as:

$$K = \frac{1+|\Delta|^2-|S_{11}|^2-|S_{22}|^2}{2|S_{21}||S_{12}|} \tag{6}$$

where

$$|\Delta| = |S_{11}S_{22}-S_{12}S_{21}|.$$

For unconditional stability, $K > 1$ and $\Delta < 1$. Thus, higher reverse isolation (S_{12}) improves the stability of an amplifier. In a common-gate LNA, since the gate of the MOSFET is conventionally connected to an AC ground, there is no Miller effect associated with the feed-forward capacitor C_{gd}. This improves the reverse isolation and hence, the stability.

3.2. **Common-source LNA**

3.2.1. *Input matching*

The common-source LNA shown in Fig. 11 uses inductive degeneration to match the input impedance, $(g_m/C_{gs})L_s$ to 50Ω [24]:

$$Z_{in} = s(L_g + L_s) + \frac{1}{sC_{gs}} + \left(\frac{g_m}{C_{gs}}\right)L_s \tag{7}$$

where L_g and L_s are chosen to resonate with C_{gs} at the operating frequency to give an inherently noiseless input match. However, the input matching network is narrowband in nature compared to its common-gate counterpart.

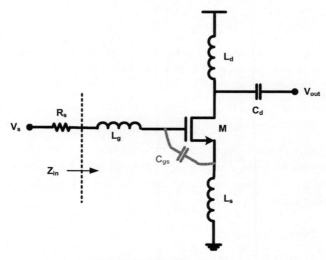

Fig. 11. Common-source LNA

3.2.2. *Noise factor*

Both the gate noise and the correlated noise contribute significantly to the noise factor of the common-source LNA, apart from the channel noise. Using a small-signal circuit for noise analysis (Fig. 12), the noise factor is given by:

$$F = \frac{\overline{i_{n,out}^2}}{\overline{i_{Rs}^2}} = 1 + \frac{\gamma}{\alpha}\frac{1}{Q}\left(\frac{\omega_0}{\omega_T}\right)\left[1 + \frac{\delta\alpha^2}{5\gamma}(1+Q^2) + 2|c|\sqrt{\frac{\delta\alpha^2}{5\gamma}}\right] \tag{8}$$

where

$$\overline{i_{Rs}^2} = \frac{4kT\Delta f}{R_s}$$

$$\overline{i_d^2} = 4kT\gamma\, g_{d0}\, \Delta f$$

$$\overline{i_g^2} = 4kT\delta\, g_g\, \Delta f$$

$$Q = \frac{1}{\omega_0\, C_{gs} R_s}$$

$$\overline{i_g i_d^*} = c\sqrt{\overline{i_g^2}\,\overline{i_d^2}}\, .$$

Here $c = j0.395$, α, γ, and δ are bias-dependent parameters [21], ω_0 is the operating frequency, and ω_T is the unity current gain frequency.

From (8), it is clear that whereas the gate noise is directly related to the Q of the input tank, the channel noise is inversely related. Thus, an optimum Q can be calculated for the minimum noise factor. The Q_{opt} and F_{min} are, respectively, given by [20]:

$$Q_{opt} = \sqrt{1 + 2|c|\sqrt{\frac{5\gamma}{\delta\alpha^2}} + \frac{5\gamma}{\delta\alpha^2}} \tag{9}$$

$$F_{min} = 1 + \frac{\gamma}{\alpha}\left(\frac{\omega_0}{\omega_T}\right)\frac{2\delta\alpha^2}{5\gamma}Q_{opt} \tag{10}$$

For low values of ω_0/ω_T, it is possible to achieve a very good noise factor in a common-source LNA, which makes it attractive for low frequency (compared to ω_T) implementations. However, it must be noted here that the above expression for noise factor does not include the contribution of noise from the series resistance of L_g. If the quality factor of L_g is low (which is usually the case for on-chip inductors), it may lead to significant degradation in the overall noise factor of the common-source LNA. Hence, in a majority of applications, L_g is implemented off-chip to maintain good noise performance.

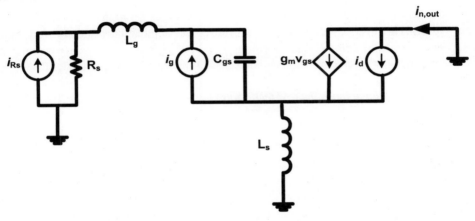

Fig. 12. Small-signal model for noise figure analysis of the common-source LNA

3.2.3. *Gain*

The effective transconductance of the common-source LNA, under perfect input matching conditions, is:

$$G_{m,eff} = g_m Q = \frac{1}{2R_S}\left(\frac{\omega_T}{\omega_0}\right) \tag{11}$$

Since ω_T/ω_0 is usually 5~10, a common-source LNA has superior gain performance to the common-gate LNA.

3.2.4. Robustness

Inductors L_s, L_g and the parasitic capacitances C_{gs} form a series resonant RLC network at the input of the common-source LNA. The quality factor of this series tank is:

$$Q = \frac{1}{2\omega_0 C_{gs} R_s} > 1 \qquad (12)$$

Thus, a higher Q makes the common-source topology less robust against PVT variations [22].

3.2.5. Reverse isolation and stability

Owing to the feed-forward capacitor C_{gd}, Miller effect exists in the common-source LNA and degrades its reverse isolation and stability [21].

To summarize, the common-source LNA is better suited for high gain and low noise applications. However, for fully integrated and robust applications, and for very high frequency designs, a common-gate topology is preferred. Next, we discuss techniques [19] to improve the reverse isolation and noise performance of a common-gate LNA.

3.3. LNA performance improvement techniques

3.3.1. Cascode

To reduce the Miller effect arising from the feed-forward capacitor C_{gd}, a common-gate NMOS is connected in cascode to the common-source NMOS (Fig. 13). This improves the reverse-isolation (S_{12}) and thus the stability of the amplifier [21]. Cascoding also helps in decoupling the input and output matching networks of the amplifier.

However, cascading also has drawbacks. The channel noise from the cascode transistor adds to the overall noise factor of the LNA. The parasitic capacitance at the common-node of the two NMOS gives another path for the signal current to flow, leading to attenuation and a higher noise factor. To reduce the parasitic capacitances, a dual-gate layout [25] can be used to combine the drain of M_1 and source of M_2 (Fig. 14).

3.3.2. Bondpad shielding

A conventional bondpad has a parasitic resistance because of the lossy Si-substrate (Fig. 15) [26]. This resistance degrades the optimum impedance match at the operating frequency and adds thermal noise, thus compromising both the noise factor and the gain of the amplifier. A possible remedy is to use a shielded bondpad that removes the effect of the parasitic resistance [27].

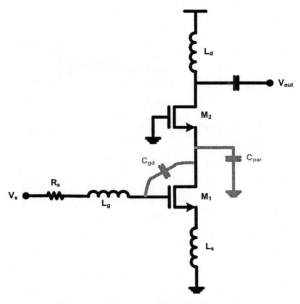

Fig. 13. Common-source LNA with cascade

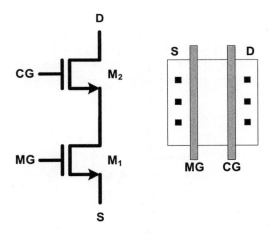

Fig. 14. Dual-gate layout

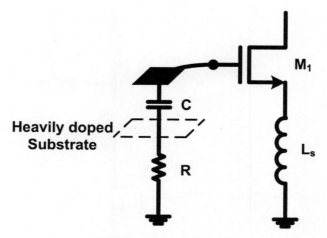

Fig. 15. Bondpad shielding

3.3.3. *Substrate contacts*

The lossy Si-substrate also leads to the parasitic resistance R_b as shown in Fig. 16 [28]. The thermal noise in R_b modulates the back-gate of the NMOS, producing "epi noise" [19]:

$$\overline{i^2_{nd,sub}} = 4kT\, R_b\, g^2_{mb}\Delta f \tag{13}$$

To reduce the effect of this substrate noise, a large number of substrate contacts can be placed near the device [29] (Fig. 17). This reduces the substrate resistance R_b and helps improve the noise performance of the amplifier.

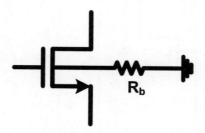

Fig. 16. Parasitic resistance R_b

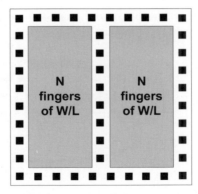

Fig. 17. Layout with a large number of substrate contacts to decrease R_b

4. Mixers

In wireless receivers, the LNA is typically followed by a number of mixers that perform frequency translation, as depicted in Fig. 18 [3]. The main purpose of this operation is to shift the RF signal to a lower frequency where processing is easier and more efficient. Mixers have two sets of inputs, an RF signal that is to be processed and an LO signal that establishes the output IF frequency as shown in Fig 19. The important performance specifications of mixers are: conversion gain, linearity, noise figure, RF, LO and IF frequency ranges, and isolation.

In the following section, different types of mixers and their applications are discussed and several innovative topologies that have emerged recently to target specific shortcomings of conventional mixers are described. In particular, two CMOS implementations are discussed in detail: A CMOS even-harmonic mixer targeting direct-conversion receivers, and an image-rejection down-converter for low-IF receivers.

4.1. *Mixers classified*

4.1.1. *Active and passive mixers*

Mixers are broadly classified into two types: passive and active [30]. Passive mixers, as the name suggests, do not dissipate dc power on their own, which is a very attractive characteristic. However, as a result, they have a negative conversion gain, and are therefore not very popular due to noise considerations. They are usually realized as a series of switches that are commutated such that the RF and LO signals are mixed together to the required IF. The switches are generally realized using MOS transistors, as shown in Fig. 13, since these are well suited for commutation switches. For low switch resistance and high conversion gain, two conditions need to be met: large switch W/L and large LO amplitude. Usually, passive mixers have better linearity performance than active implementations, but this linearity-gain tradeoff is not enough to offset the noise figure problems associated with these mixers.

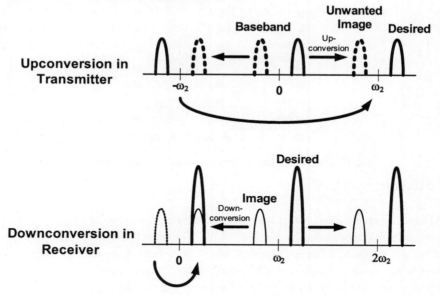

Fig. 18. Frequency translation in transmitter and receiver

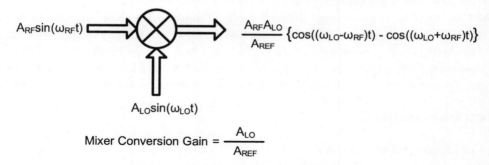

Fig. 19. Mixer with RF and LO inputs and an IF output

Active mixers, on the other hand, do dissipate dc power, and have a net conversion gain that can be set by the designer. This allows a better transmit/receive chain optimization, and more degrees of freedom to tradeoff power, noise and linearity. They usually have two distinct sections: a transconductance stage and a frequency translation stage. The transconductance stage at the RF input converts the incoming voltage signal into a current signal. This current signal is passed through the frequency translation stage that is usually in series with the RF stage (i.e., the same current passes through them). The frequency translation stage mainly consists of a set of transistors that are driven at the LO frequency and therefore switch at the same speed. The net effect is to multiply the two signals to produce a down-converted signal. The LO amplitude required is only a few hundred millivolts (set by linearity considerations). Active mixers thus have many advantages over passive ones that make them a good choice for RF implementations. However, this does come with a price: Active mixers usually have more devices than passive ones, which directly translates to more noise in the mixer output. In particular, flicker noise has arisen as a significant problem in integrated CMOS RF mixers.

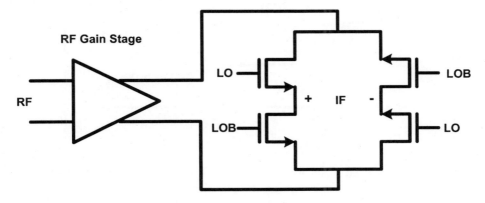

Fig. 20. Passive CMOS mixer

The Gilbert Mixer, shown in Fig 21, has become ubiquitous in modern integrated wireless communication systems. It offers good conversion gain, isolation and a good tradeoff between linearity, noise and power consumption [31]. The original Gilbert Mixer was BJT-based, but is easily imported into CMOS implementations where it is well suited for integration with other CMOS circuitry. Since the loads are usually resistive in a down-conversion mixer, the overall noise is set by the noise of the V-I converter (transconductance stage) and the load resistors. The overall linearity performance is limited to that of the V-I converter.

4.1.2. *Single- and double-balanced mixers*

One major concern in RF mixers is isolation from the RF and LO ports to the output IF port. It is desirable to have minimum feedthrough of RF and LO signals to the output. Theoretically, a series combination of two MOSFETs as represented in Fig. 22, can serve as a mixer, with one being the V-I stage, and the other being the switching stage. However, it has both LO and RF components intermixed with the IF component in the output signal waveform. The RF component can be removed from the output by feeding the LO waveform in a differential manner to the switching stage, and taking the output also as a differential signal. This is called a single-balanced mixer, and is depicted in Fig. 23. This structure still has LO feedthrough to the output IF port. Consequently, the topology can be further modified to cancel the LO feedthrough. This is accomplished by differentially driving the RF port, as shown in the double-balanced mixer configuration in Fig. 21.

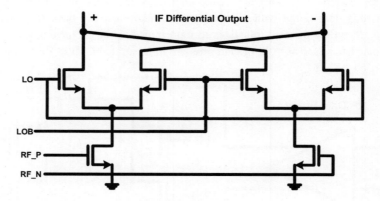

Fig. 21. Active CMOS mixer—Gilbert cell

The double-balanced mixer is extensively used in most wireless integrated systems in both receive and transmit modes. It can further be used in an image-reject configuration to eliminate the image problem in RF receivers, as in the Hartley architecture shown in Fig. 24. This architecture exploits the opposite phases of the signal and image when mixed with LO signals that are in quadrature [32].

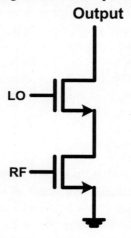

Fig. 22. Single-ended active mixer

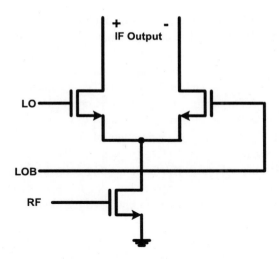

Fig. 23. Single-balanced Gilbert mixer

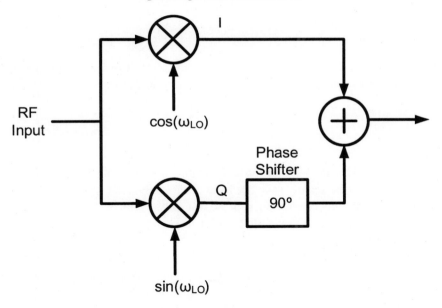

Fig. 24. Hartley architecture

4.2. *Harmonic mixers*

The direct-conversion receiver architecture has gained substantial attention because it enables a high level of integration. However, the problems of LO self-mixing (DC offset), IM2 distortion and 1/f noise have remained serious impediments to its widespread use in integrated receiver systems. To ease some of these difficulties, AC coupling capacitors or feedback circuitry has commonly been used. The AC coupling solution suffers from large capacitor values and long settling times, while feedback is complex and requires a wide feedback loop bandwidth.

As a more feasible approach, even-harmonic mixers have been proposed previously using anti-parallel diode-pairs, emitter-coupled BJTs and more recently, in a digital-CMOS technology [33]. The single-balanced version of the CMOS even-harmonic mixer (EHM) is shown in Fig. 25, and the corresponding LO waveforms are described in Fig 26. These example figures show the operation of a second-harmonic EHM. These topologies work by separating the operating frequencies of the LO and RF, while keeping the effective LO the same as before. The fundamental mixing between RF and LO signals is suppressed, and at the same time, mixing between RF and even harmonics of LO is allowed. The LO self-mixing mechanisms in both a conventional mixer and an EHM are shown in Fig.27.

Because a square wave contains no even-harmonics, mixing in a conventional mixer takes place with odd harmonics of the LO signal only (the fundamental being the most desired component).

$$v_{LO}(t) = \frac{4}{\pi}\left(\cos\omega_0 t - \frac{1}{3}\cos 3\omega_0 t + \frac{1}{5}\cos 5\omega_0 t - \frac{1}{7}\cos 7\omega_0 t + ...\right) \tag{14}$$

To increase the even-harmonic content of the square wave, it is necessary to change the duty-cycle of the waveform. The frequency content of a square waveform with duty cycle d is given by

$$v_{LO}(t) = \frac{4}{\pi}\left(\sin d\pi \cos\omega_0 t + \frac{1}{2}\sin 2d\pi \cos 2\omega_0 t + ... + \frac{1}{n}\sin nd\pi \cos n\omega_0 t\right) \tag{15}$$

At exactly 0 and 50% duty-cycles, no even harmonics exist and the maximum even-harmonic content in the square wave occurs at a 25% duty cycle. As a result, the mixer would realize maximum conversion gain at this duty cycle of 25%. To prevent LO feedthrough to the output, a double-balanced version of the even-harmonic mixer can also be implemented and is depicted in Fig 28 [34].

Considering it from a different perspective, the simple MOSFET switch in the LO stage of the conventional mixer has been replaced by a complex switch. The complex (digital) switch has multi-phase inputs that operate at a different frequency, but the overall current commutation occurs at the desired frequency that is usually an even harmonic of the actual LO frequency. It is possible to architecturally simplify the mixer even more, by digitally combining the multi-phase LO signals to create the effective-LO signal outside the main mixer. This can then be applied to a conventional Gilbert-cell mixer. However, this would defeat the purpose of the EHM itself, which is to avoid direct introduction of any LO signals in the RF band that could couple through to the mixer RF input.

Any mismatch between MOSFET pairs can cause residual LO and RF feedthrough to the output, and also reduce DC offset cancellation. Careful layout of the transistors is required to create well-matched devices. With typical device mismatches taken into consideration, this architecture exhibits excellent DC offset cancellation. Moreover, this DC offset cancellation performance is quite insensitive to input power level and LO phase mismatches.

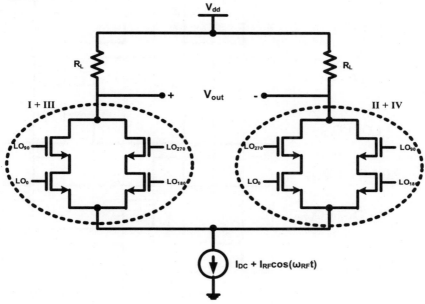

Fig. 25. CMOS single-balanced even-harmonic mixer

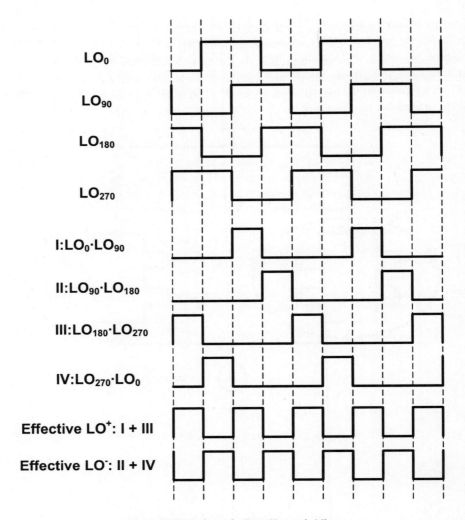

Fig. 26. LO Waveforms for Even-Harmonic Mixer

4.3. *Image-reject down-converter for low-IF receivers*

As mentioned previously, DC offset, IM2 distortion and 1/f noise remain serious hurdles to the wide use of DCR front-ends in CMOS-based integrated receiver systems. Low-IF receivers avoid many of these issues, while at the same time, achieving similar levels of integration. DC offset and flicker noise concerns are eliminated locating the low-IF frequency band just above the flicker noise corner. However, this implementation is quite challenging due to the associated stringent image-rejection (IR) requirements. High bandwidth at low-IF increases the bandwidth ratio and aggravates phase errors in the 90°

phase difference network. This, in addition to the inherent phase errors due to gain and phase mismatches, restricts the achievable IR ratio [34].

From the above considerations, it is evident that careful selection of the IF frequency is paramount. By choosing the IF frequency equal to half the channel spacing, the image signal can be placed in the adjacent channel, and the IR ratio reduced to the adjacent channel selectivity. A dual conversion architecture as shown in Fig. 29, is appropriate, with the second IF placed at $f_{ch}/2$ (half the channel spacing). The duplexer isolates the transmit and receive bands in frequency-division duplexing (FDD) systems, while the SAW filter suppresses out-of-band blockers in addition to providing image rejection for the first mixer. The down-converter consists of two mixing stages as shown in Fig. 30. The first stage may be implemented as a conversion Gilbert-cell mixer. The second stage is implemented as a *masking quadrature mixer* (MQM) that is insensitive to amplitude and phase variations in the LO signal. The MQM down-converts the desired signal to the low-IF band at $f_{ch}/2$ [34].

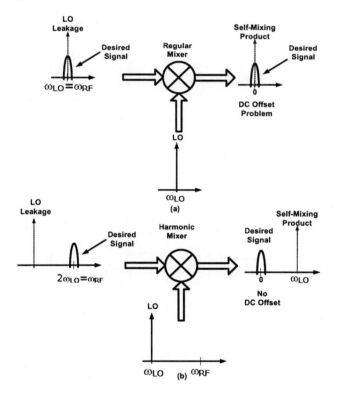

Fig. 27. LO self-mixing mechanisms in (a) conventional, and (b) EHM mixers

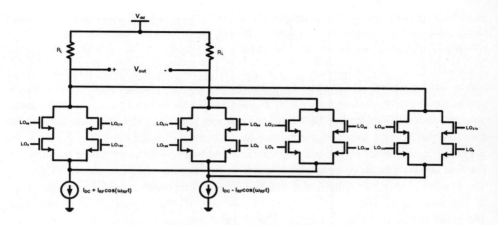

Fig. 28. CMOS double-balanced even-harmonic mixer

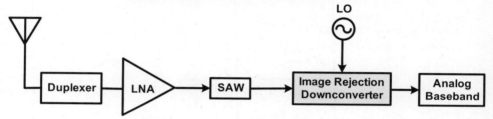

Fig. 29. Dual-conversion low-IF receiver architecture

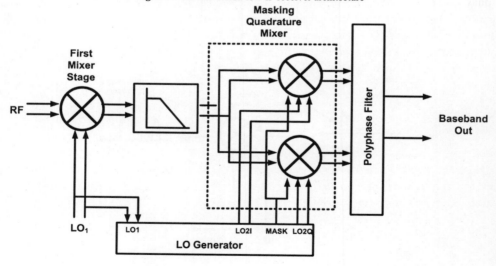

Fig. 30. Masking quadrature mixer

The MQM exhibits minimal amplitude errors since it has a single transconductance stage, and the LO switches are driven by rail-to-rail digital signals. Higher clock frequencies are used to derive the LO signals, and digital LO correction is employed, both leading to accurate quadrature phase delays. For example, if there are two signals at frequencies f_{LO} and $4f_{LO}$, then the period of the second signal is exactly 90° of that of the first signal. Similarly, the rising and falling edges of a clock running at $2f_{LO}$ are 90° apart, as shown in Fig. 24. Therefore, the LO generator mixes the LO signal with the faster pulses to produce precise in-phase and quadrature IF signals. The mixer conversion gain is $\dfrac{2}{\pi} g_m R_L \cdot \sin d\pi$ where d is the fractional duty cycle of the LO signal, and g_m and R_L are the transconductance and load resistance, respectively. Nevertheless, any mismatches in the uncorrected LO signals are directly manifested as gain and phase mismatches in the quadrature LO signals. This issue is resolved by performing an AND operation between the uncorrected LO signals and center-aligned $2f_{LO}$ signals to generate LO signals whose quadrature accuracy is ensured.

In spite of this, a non-50% duty cycle in the $2f_{LO}$ signals also causes quadrature phase and pulse-width errors. This problem is overcome by using yet another signal of $8f_{LO}$ that has an accurate period. We can then take advantage of the accurate edges of this clock to *mask* the edges of the generated quadrature LO signals, as shown in Fig. 25. Thus, the allowable duty cycle errors of 50% ± 6.25% are allowed in the LO signal. This comes at a small price of slightly reduced mixer gain due to duty cycle (d) reduction. If more loss cannot be tolerated, then higher frequencies can be used to mask the edges at the expense of smaller duty cycle error correction.

An essential component of the image-rejection down-converter is a wideband 90° phase-difference network. A multi-stage RC poly-phase filter is a natural choice since its passive nature consumes less current and introduces fewer noise sources than an active one [34][35]. Its potential uses in transceiver systems include single-sideband modulation, quadrature carrier generation, and image-rejection. The poly-phase filter is a sequence-asymmetric system that selectively passes one sequence of signals while rejecting another. Thus, the network can be arranged such that it exhibits a pass-transfer-function for the desired RF signal, while rejecting the unwanted image signal.

A single RC poly-phase filter stage achieves image-rejection at a single frequency value. Cascading multiple filter stages and thus staggering the gain magnitude minima of successive stages, wideband image-rejection can be achieved. Usually, a large number of stages are required when the bandwidth ratio f_H/f_L is high and the center frequency is low, or to accommodate process and temperature variations of on-chip passive elements as well as device parasitics.

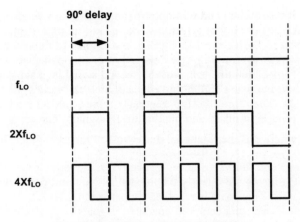

Fig. 31. Phase relationships between clocks

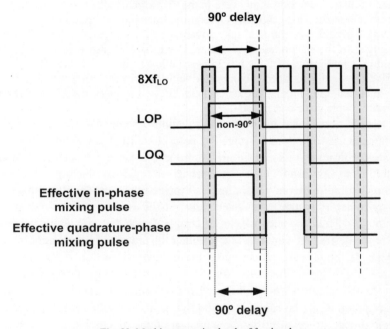

Fig. 32. Masking operation by the *8f_{LO}* signals

5.   Voltage-controlled Oscillators

The LO port of the mixer is driven by a phase-locked loop (PLL) frequency synthesizer. Shown in Fig. 33, a typical PLL consists of a phase detector, a charge-pump, a loop filter, a voltage-controlled oscillator (VCO), and a frequency divider connected in a loop so its output frequency can be locked and will not drift in the presence of device noise. One of the most important blocks in the PLL is the voltage-controlled oscillator.

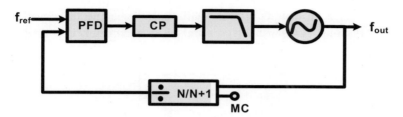

Fig. 33. Typical phase-locked loop in wireless transceivers

Oscillators can be classified into two categories: relaxation oscillators and harmonic oscillators. Relaxation oscillators usually have high harmonic content and poor phase noise performance while harmonic oscillators, which typically use *LC* resonators or crystals, have high spectral purity and offer good noise performance. The majority of the oscillators used in wireless applications today are resonator-based oscillators. Shown in Fig. 34, the cross-coupled LC VCO is arguably the most common VCO being designed in the world. The VCO can be modeled as shown in Fig. 35. The inductor *L*, varactors *Cvar*, and parasitics of the MOSFETs form a resonant network that defines the frequency of oscillation:

$$\omega_0 = \frac{1}{\sqrt{LC}} \tag{16}$$

The parallel resistance *R* models the loss due to the metal resistance in the inductor and the varactor. The NMOS and PMOS cross-coupled pairs form a small-signal negative resistance generator to cancel this loss resistance so that oscillation can be sustained. This condition determines the minimum device transconductance for oscillation to start:

$$\frac{1}{R} - \frac{gm_p + gm_n}{2} \leq 0 \tag{17a}$$

$$gm_p + gm_n > \frac{2}{R} \tag{17b}$$

The two most important merits that characterize a VCO are phase noise and tuning range. Phase noise is the short-term frequency instability caused by noise in the devices and is defined as [37]:

$$L(\Delta\omega) = 10 \cdot \log\left[\frac{P_{sideband}(\omega_0 + \Delta\omega, 1Hz)}{P_{sig}}\right] \tag{18}$$

where $P_{sideband}(\omega_0 + \Delta\omega, 1Hz)$ is the single sideband power at an offset frequency from the carrier in a 1Hz bandwidth, and P_{sig} is the carrier power. The units of phase noise are decibels below the carrier per Hertz. The harmful effect of phase noise can be illustrated through a phenomenon called reciprocal mixing. Consider the down-conversion operation on a desired signal that is close to a strong signal in the frequency spectrum shown in Fig. 36. Part of this interfering signal will be down-converted to the intermediate frequency due to the sidebands in the LO signal, significantly degrading the sensitivity of the system. Thus, ensuring VCOs with high spectral purity is extremely important.

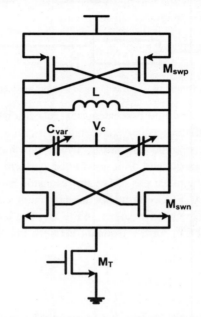

Figure 34. Cross-coupled LC VCO

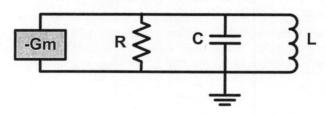

Figure 35. LC VCO model

One of the most challenging tasks in designing a VCO is the modeling of the phase noise. In the past several decades, significant research efforts have yielded a number of phase noise models: a linear time-invariant model [35], a linear time-variant model [37], a non-linear model [38], and a physical model [39]. Among them, the linear time-variant model is the most widely recognized by the IC design community:

$$\text{phase noise} \equiv L\{\Delta\omega\} = 10\log\left(\frac{\overline{i_n^2}/\Delta f}{2\cdot\Delta\omega^2}\cdot\frac{\Gamma_{eff-rms}^2}{Q_{max}^2}\right) \tag{19}$$

where $\Delta\omega$ is the offset frequency from the carrier, $\overline{i_n^2}/\Delta f$ is the power spectral density of current noise source, Q_{max} is the maximum charge swing, and $\Gamma_{eff-rms}$ is the root-mean-square value of the effective impulse sensitivity function (ISF). The ISF, which is a periodic dimensionless quantity that describes sensitivity of an oscillator to an impulsive input, can be obtained directly by injecting a noise current into the VCO output and observing its phase shift. To minimize phase noise, the ISF and Q_{max} need to be minimized and maximized, respectively.

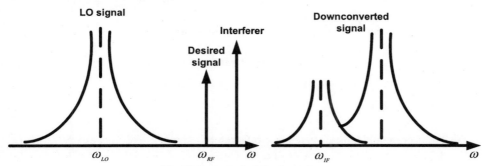

Fig. 36. Reciprocal mixing in receivers

The tuning range of the VCO, which is defined as ω_{max}-ω_{min}, where ω_{max} and ω_{min} are the maximum and minimum oscillation frequencies, respectively, is primarily determined by the maximum to minimum capacitance ratio of the varactor, C_{var}. Accumulation-mode MOS varactors are most commonly used for their wide linear ranges, and a typical C-V plot is shown in Fig. 37. While wide tuning range is desirable because it offers high system capacity in a transceiver, it also offers high VCO gain, which is defined as the ratio of the tuning range and the range of the control voltage (Vc). A large VCO gain makes the VCO sensitive to noise on the control voltage and degrades the phase noise.

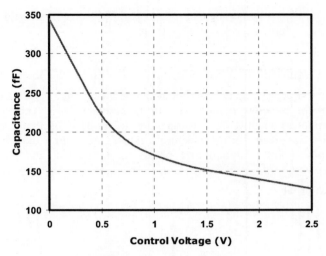

Fig. 37. Typical C-V plot of a varactor

5.1. *Quadrature VCO*

In all communication systems, quadrature signals are necessary for I/Q (de)modulation. Thus, it is important for VCOs to provide in-phase and quadrature LO signals. As stated in Section II, several options exist for on-chip quadrature carrier generation. One approach involves feeding a single VCO output to a poly-phase filter, which in its simplest form is an RC-CR passive network. The major shortcomings of this method are excessive die area and quadrature phase mismatches that strongly depend on device matching and operating frequency range. A second alternative is to combine a single VCO running at twice the frequency of interest with two divide-by-two flip-flops. This technique suffers from increased power consumption and sensitivity to the duty cycle of the VCO waveform. The design of a VCO operating at twice the original frequency is also more difficult. The third option, anti-phase injection, achieves low phase noise and high quadrature phase accuracy [40][41]. Its primary drawback is doubled die area and power consumption compared to a single VCO. The fundamental principle behind anti-phase injection is illustrated below in Fig. 38. The outputs of the oscillator on the left (V_{out2}: A and B) are anti-phase coupled to the oscillator on the right. The outputs of the oscillator on the right (V_{out1}: C and D) are in-phase coupled to that on the left. The two oscillators can be modeled as shown in Fig. 39, and the two outputs can be written as:

$$V_{out1} = G_{m2}V_{out2} \cdot \frac{-RZ}{Z-R}$$
$$V_{out2} = G_{m1}V_{out1} \cdot \frac{-RZ}{Z-R}$$

(20)

Assuming that $G_{m1}=G_{m2}$, rearranging and solving the above equations yields:

$$V_{out\,1} = \pm jV_{out\,2} \tag{21}$$

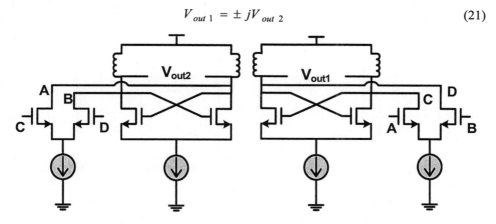

Fig. 38. Coupling of two VCOs to create in-phase and quadrature signals

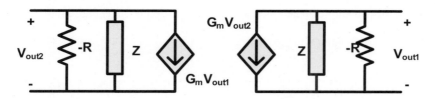

Fig. 39. Small-signal equivalent model of QVCO

One of key issues in quadrature VCO (QVCO) design is the placement of the injection transistors and their impact on various performance parameters of the oscillator. When the monolithic QVCO was first proposed, the injection transistors were connected in parallel with the –Gm devices, similar to what is shown in Fig. 38. Due to the noise of the injection transistors and degraded Q-factor of the tank, the phase noise of the QVCO is severely degraded. In addition, the phase error depends on the size ratio of the switching and injection transistors. Instead of connecting in parallel with the –Gm devices, the injection transistors can be connected in series. A QVCO with cascode injection transistors (CI-QVCO), shown in Fig. 40 does not need the phase error tradeoff to achieve good phase noise performance. However, such optimized performance is only achieved by sizing the injection transistors five times wider than the –*Gm* devices; this renders the CI-QVCO topology less attractive for high frequency and large tuning range applications. Instead of employing the injection transistors as cascode devices, they can be positioned below the –*Gm* transistors as shown in Fig. 41. This degeneration-injected topology (DI-QVCO) is motivated by the fact that the wide M_{cp1} transistors in Fig. 40 significantly reduce the VCO tuning range by contributing a large amount of parasitic capacitance at the output nodes. If the size of M_{cp1} is reduced to increase the tuning range, the start-up loop gain factor is significantly reduced because M_{sw1} is forced to operate in the deep triode region. The start-up loop gain expressions of the CI-QVCO and the DI-QVCO topologies can be derived and compared. The loop gain of the CI-QVCO is the multiplicative gain of two single-stage common-source cascode amplifiers with tuned loads:

$$T_{CI-QVCO} \approx -(g_{m-sw1} \cdot r_{ds-sw1})^2 (g_{m-cp1} \cdot r_{ds-cp1})^2 \cdot$$

$$\left[\frac{j\omega L}{j\omega L + (g_{m-cp1} r_{ds-cp1} r_{ds-sw1})(1 - \omega^2 LC + j\omega LG)} \right]^2 \tag{22}$$

where g_{m-sw1}, g_{m-cp1}, r_{ds-sw1}, and r_{ds-cp1} are the small-signal transconductances and resistances of the –Gm and injection transistors, respectively, G is the conductance of the tuned load, and T is the loop gain. For simplicity, parasitic capacitance at the output is lumped into the tuned load and that at node X is neglected.

For the DI-QVCO, however, the -Gm transistors M_{sw2} operate in saturation and the injection transistors M_{cp2} operate in the triode region. The loop gain now consists of the gain product of two single-stage tuned amplifiers with active degeneration:

$$T_{DI-QVCO} \approx -(g_{m-sw2} \cdot r_{ds-sw2})^2$$

$$\cdot \left[\frac{j\omega L}{j\omega L + (g_{m-sw2} r_{ds-cp2} r_{ds-sw2})(1 - \omega^2 LC + j\omega LG)} \right]^2 \tag{23}$$

Reference [38] asserts that sizing the injection transistors about five times larger than the –Gm transistors in the CI-QVCO topology yields optimum performance. Inspection of the CI-QVCO loop gain expression reveals that maintaining a large ratio α defined as:

$$\alpha = \frac{W_{cp1}}{W_{sw1}} \tag{24}$$

is necessary for reliable oscillation start up; keeping α large causes the -Gm transistor M_{sw1} to operate near the transistor pinch-off point instead of deep in the triode region. Hence, if α decreases the voltage at node X decreases, which drives M_{sw1} deep into triode operation and reduces the loop gain. Consequently, the large size of the injection transistors becomes a design constraint. The start-up for DI-QVCO, on the other hand, is not substantially impacted by the ratio α because the injection transistor M_{cp2}, operating in the triode region, merely acts as an active degeneration device and does not play a significant role in defining the loop gain.

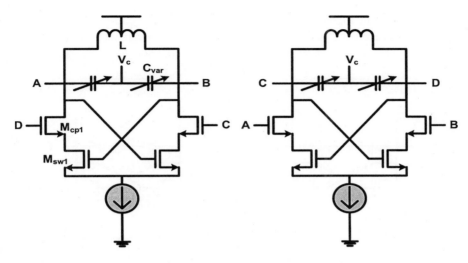

Fig. 40. QVCO with cascode injection transistors

The start-up loop gain analysis above leads to another interesting observation. From [37], we know that the complementary VCO topology is superior to its PMOS- or NMOS-only counterparts because of its larger output swing for the same power dissipation and its rise- and fall-time symmetries. Due to the start-up constraint on α, the complementary version of CI-QVCO is impractical where high frequency signals and large tuning range are required. The DI-QVCO on the other hand, does not have such a limitation. It can be seen in Fig. 42 that an optimized NMOS-only DI-QVCO at 5GHz has nearly three times the tuning range as its cascode-injected counterpart.

To make a fair comparison of the phase noise performances between the two topologies, each is optimized for the same design specifications at 5GHz using a total of 1600 simulation runs. The quality factors of the tanks and the power consumptions are kept equal [43]. The ISF and the phase noise of the two topologies are shown in Figs. 43 and 44, respectively. The ISF of the CI-QVCO displays nearly twice the amplitude as DI-QVCO, implying that the DI-QVCO offers superior phase noise performance. From Fig. 44, DI-QVCO outperforms CI-QVCO by 8dB and 4dB at close-in and far-out frequencies, respectively.

Unlike QVCOs with parallel injection transistors wherein quadrature phase error is strongly dependent on α, phase error is independent of α in series-coupled QVCOs [41]. Thus, quadrature phase error for CI-QVCO and DI-QVCO depends largely on device mismatches. To quantify phase error performance, 1000 simulations including device mismatches were performed on each optimized design. Mismatches in tank components and device geometries of injection and −Gm transistors are swept individually from σ =1% to σ =5%. Uniform 10% variations in the threshold voltages and power supply voltage are also assumed. Temperature is randomly varied from 0 to 75°C. Figures 45 through 47 show the standard deviations of quadrature phase error versus device mismatches for both topologies. It can be seen that mismatches in coupling devices significantly impacts the CI-QVCO while the phase error performance of the DI-QVCO is more susceptible to mismatches in switching devices and tank components.

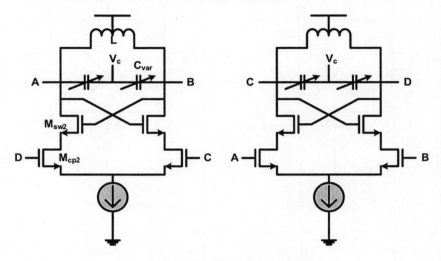

Fig. 41. QVCO with degenerated injection transistors

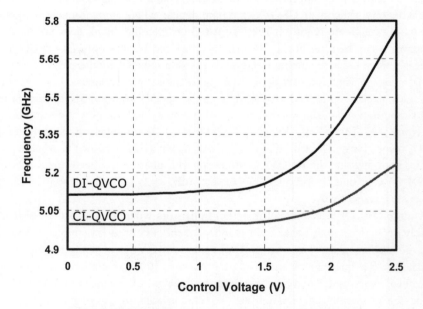

Fig. 42. Tuning range comparison of DI-QVCO and CI-QVCO

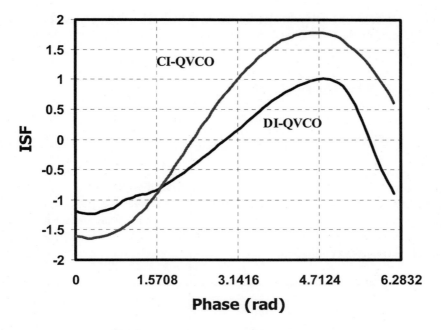

Fig. 43. ISF comparison of the DI-QVCO and CI-QVCO.

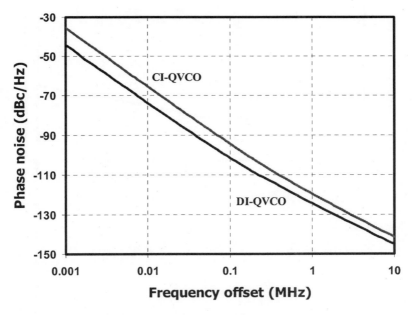

Fig. 44. Phase noise comparison of the DI-QVCO and CI-QVCO

Table 1 compares the optimized performance of both QVCO topologies at 5GHz. The VCO figure-of-merit (FOM) and power-frequency-tuning-normalized figure-of-merit (PFTN-FOM) are defined as:

$$FOM = 10 \cdot \log_{10}\left[\left(\frac{\omega_0}{\Delta\omega}\right)^2 \cdot \frac{1}{L\{\Delta\omega\} \cdot P_{diss}\big|_{mW}}\right] \tag{25}$$

$$\text{PFTN} - \text{FOM} = 10 \cdot \log\left[\frac{kT}{P}\left(\frac{\omega_{tune}}{\Delta\omega}\right)^2 \frac{1}{L\{\Delta\omega\}}\right] \tag{26}$$

where ω_0 is the center frequency, P is the power dissipation in Watts, k is the Boltzmann's constant, T is the absolute temperature in degrees Kelvin, ω_{tune} is the tuning frequency range, and $\Delta\omega$ and $L\{\Delta\omega\}$ are the offset frequency with respect to the carrier and its associated phase noise [43]. Clearly, CI-QVCO is outperformed by its degeneration-injected counterpart.

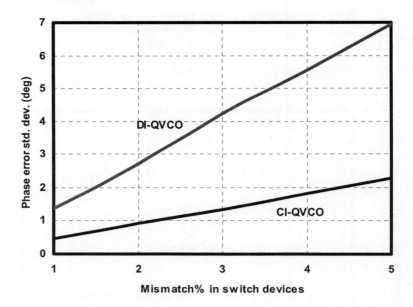

Fig. 45. Phase error standard deviation comparison between DI-QVCO and CI-QVCO where mismatches exist in the switch devices

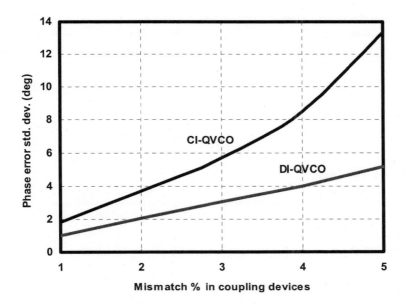

Fig. 46. Phase error standard deviation comparison between DI-QVCO and CI-QVCO where mismatches exist in coupling devices

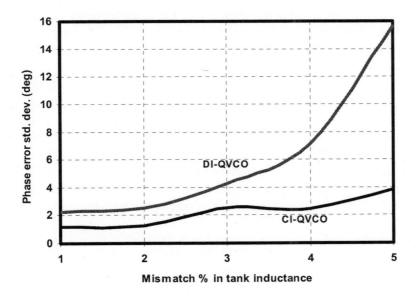

Fig. 47. Phase error standard deviation comparison between DI-QVCO and CI-QVCO where mismatches exist in the tank inductances

Table 1: CI-QVCO and DI-QVCO performance comparison

	CI-QVCO	DI-QVCO
ω_0	5.15GHz	5.4GHz
PN @10kHz	-65.3dBc/Hz	-73.74dBc/Hz
PN @ 3MHz	-130.38dBc/Hz	-134.4dBc/Hz
Tuning Range	230MHz	650MHz
Phase Error	0.085°	0.024°
Power Supply	2.5V	2.5V
Power	20mW	20mW
FOM	181.6dB	186.5dB
PTFN-FOM	-19.3dB	-6.3dB

5.2. *Colpitts VCO*

While cross-coupled LC VCOs are popular because of their good start-up condition and ease of implementation, they have one major drawback: noise generation by the -Gm devices occurs at a moment when the oscillator is most sensitive to perturbations, which degrades the phase noise considerably. Colpitts oscillators, on the other hand, can alleviate this issue and achieve low phase noise because noise injection in Colpitts oscillators does not occur at the zero crossings of the output waveform. Fig. 48 shows a single-ended Colpitts oscillator. By calculating the impedance of the small-signal model, the minimum device transconductance for oscillation startup can be found:

$$gm \geq gm_c = \omega_0^2 C_1 C_2 R_s \qquad (27)$$

where ω_0 is the oscillating frequency and R_s is the series resistance of the inductor. Examining the above inequality reveals a drawback of the Colpitts oscillator; that is, the device transconductance required for start-up is proportional to C_1 and C_2. In state-of-the-art CMOS technologies where process variations can be as much as 20%, start-up reliability is difficult to guarantee if the oscillator is to be operated at high frequencies where the parasitic capacitance significantly increases the values of C_1 and C_2. One way to overcome this start-up unreliability is to design for a large margin between the actual gm and required transconductance gm_c. This, of course, increases the power dissipation of the circuit.

The Colpitts topology can be designed in a differential form as well as used to implement quadrature VCOs [44]. Shown in Fig. 49 is one such implementation at 6GHz. Each VCO is implemented as a differential Colpitts oscillator. The outputs of the QVCO are forced to oscillate 90° out of phase using the anti-phase injection described previously. The cross-coupled devices M_{sw3} act as switching current sources while the injection devices M_{cp3} are inserted into the circuit as degeneration devices. The capacitor C_2 in Fig. 49 is combined with the varactors C_{var}. To minimize phase mismatches between the in-phase and quadrature outputs, the common-source nodes of the two VCOs are connected together. Tail resistor R_T is used for biasing instead of a MOS current source to reduce flicker noise up-conversion; inductor L_T is inserted in series with R_T to provide high impedance at the second harmonic frequency.

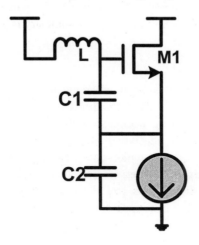

Fig. 48. A single-ended Colpitts oscillator

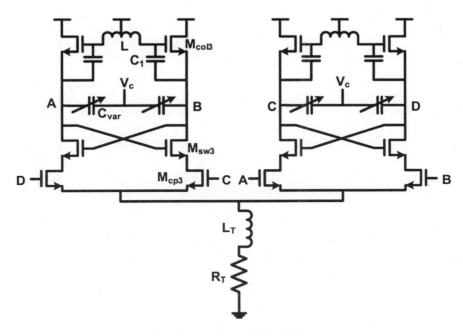

Fig. 49. A quadrature Colpitts VCO

For comparison purposes, the ISF of the 6GHz quadrature Colpitts VCO shown above is plotted against that of a 6GHz DI-QVCO in Fig. 50. It is evident that the quadrature Colpitts QVCO exhibits a smaller ISF, thus indicating its superior noise characteristics. The phase noise and tuning plots of this quadrature Colpitts VCO are shown in Fig. 51 and 52, respectively. Binary-weighted capacitor arrays are implemented to reduce the VCO gain. The overall performance summary is given in Table 2.

6. Conclusions

Several state-of-the-art wireless receiver architectures including the traditional super-heterodyne, the image reject heterodyne, the direct-conversion, and the very-low IF have been presented in this paper along with case studies. In addition, traditional as well as emerging circuit topologies for receiver building blocks have been examined. These include the common-source and common-gate LNAs, passive, active, even-harmonic, and masking quadrature mixers, and series-injected and Colpitts quadrature VCOs. With the current aggressive scaling trends and the reduced power supply voltages, these developments will be crucial in helping to achieve an integrated solution with low power and high performance.

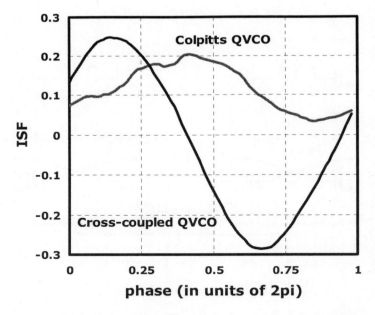

Fig. 50. ISF comparison of the Colpitts and conventional cross-coupled QVCO.

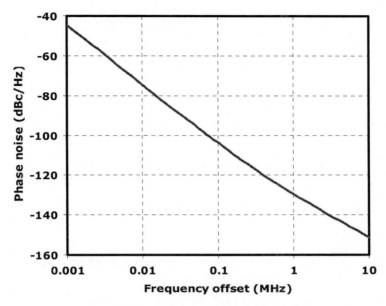

Fig. 51. Phase noise plot of a 6GHz Colpitts QVCO

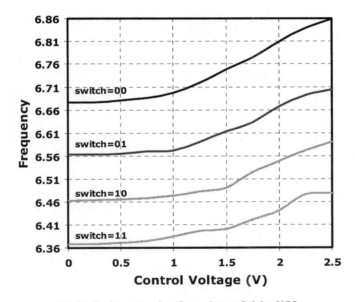

Fig. 52. Tuning range plot of a quadrature Colpitts VCO

Table 2: Colpitts QVCO performance summary

Technology	250nm SiGe BiCMOS
Center Frequency	6.5GHz
Tuning Range	500MHz
Phase Noise	-140.5dBc/Hz@3MHz
Power	20mW
Phase Error	0.55°
FOM	194dB
PFTN-FOM	-2dB

References

1. E. H. Armstrong, "A new system of short wave amplification," *Proc. IRE*, vol. 9, pp. 3-27, Feb. 1921.
2. T. Stetzler, et al., "A 2.7 V–4.5 V single-chip GSM transceiver RF integrated circuit," *IEEE J. Solid-State Circuits*, vol. 30, pp. 1421-1429, Dec. 1995.
3. B. Razavi, *RF Microelectronics,* Prentice-Hall, Englewood Cliffs, NJ, 1998.
4. D. K. Weaver, "A third method of generation and detection of single sideband signals," *Proc. IRE,* vol. 44, pp. 1703-1705, Dec. 1956.
5. M. J. Gingell, "Single sideband modulation using sequence asymmetric polyphase networks," *Electrical Communication*, vol. 48, pp. 21-25, 1973.
6. B. Razavi, "A 5.2-GHz CMOS receiver with 62-dB image rejection," *IEEE J. of Solid-State Circuits*, vol. 35, pp. 810-815, May 2001.
7. J. C. Rudell, et al., "A 1.9MHz wide-band IF double conversion CMOS receiver for cordless telephone applications," *IEEE J. of Solid State Circuits,* vol. 32, pp. 2071-2088 Dec. 1997.
8. A. Molnar, et al., "A single-chip quad band (850/900/1800/1900MHz) direct-conversion GSM/GPRS RF transceiver with integrated VCOs and fractional-N synthesizer," *IEEE International Solid-State Circuits Conference Dig. Tech. Papers*, pp. 232-233, Feb. 2002.
9. D. Haspeslagh, et al., "BBTRX: A baseband transceiver for a zero IF GSM hand portable station," *Proc. IEEE Custom Integrated Circuits Conference,* pp. 10.7.1-10.7.4, May 1992.
10. A. Behzad, et al., "Direct-conversion CMOS transceiver with automatic frequency control for 802.11a wireless LANs," *IEEE International Solid-State Circuits Conference Dig. Tech. Papers*, pp. 356-357, Feb. 2003.
11. P. Zhang, "A 5-GHz direct-conversion CMOS transceiver," *IEEE J. Solid-State Circuits*, vol. 38, pp. 2232-2238, Dec. 2003.
12. C. Enz, et al., "Circuit techniques for reducing the effects of op-amp imperfections: autozeroing, correlated double sampling, and chopper stabilization," *Proc. IEEE*, vol. 84, pp. 1584-1614, Nov. 1996.
13. I. Vassiliou, et al., "A single-chip digitally calibrated 5.15-5.825-GHz 0.18-μm CMOS transceiver for 802.11a wireless LAN," *IEEE J. Solid-State Circuits*, vol. 38, pp. 2221-2231, Dec. 2003.
14. J. Crols, et al., "Low-IF topology for high performance analog front ends of fully integrated receivers," *IEEE Trans. on Circuits and Systems-II: Analog and Digital Signal Processing,* vol. 45, pp. 269-282, Mar. 1998.
15. S. Dow, et al., "A dual-band direct-conversion/VLIF transceiver for 50GSM/GSM/DCS/PCS," *IEEE International Solid-State Circuits Conference Dig. Tech. Papers*, pp. 230-231, Feb. 2002.

16. O. Oliaei, et al., "A 5 mW ΔΣ modulator with 84dB dynamic range for GSM/EDGE," *IEEE International Solid-State Circuits Conference Dig. Tech. Papers*, pp. 46-47, Feb. 2001.

17. E. Duvivier, et al., "A fully integrated zero-IF transceiver for GSM-GPRS quad-band applications," *IEEE J. Solid-State Circuits*, vol. 38, pp. 2249-2257, Dec. 2003.

18. H. T. Friis, "Noise figures of radio receivers," *Proc. IRE*, vol. 32, pp. 419-422, July 1944.

19. D. J. Allstot, "Low noise amplifiers," *IEEE International Solid-State Circuits Conference Short Course*, Feb. 2001.

20. D. J. Allstot, et al., "Design considerations for CMOS low-noise amplifiers," *IEEE Radio Frequency Integrated Circuits (RFIC) Symposium*, pp. 97-100, June 2004.

21. Y. Cheng, et al., "High-frequency small-signal ac and noise modeling of MOSFETs for RF IC design," *IEEE Trans. Electron Devices*, vol. 49, pp. 400-408, Mar. 2002.

22. Q. Huang, et al., "The impact of scaling down to deep submicron on CMOS RF circuits," *IEEE J. Solid-State Circuits*, vol. 33, pp. 1023-1036, July 1998.

23. K. Kurokawa, "Power waves and the scattering matrix," *IEEE Trans. Microwave Theory and Tech.*, vol. MTT-13, pp. 194-202, Mar. 1965.

24. D. K. Shaeffer, et al., "A 1.5-V, 1.5-GHz CMOS low-noise amplifier," *IEEE J. Solid-State Circuits*, vol. 32, pp. 745-759, May 1997.

25. F. Stubbe, et al., "A CMOS RF-receiver front-end for 1 GHz applications," *VLSI Circuits Symposium Digest of Technical Papers*, pp. 80-83, June 1998.

26. N. Camilleri, et al., "Bonding pad models for silicon VLSI technologies and their effects on the noise figure of RF NPNs," *IEEE MTT-S Microwave Symposium Digest*, vol. 2, pp. 1179-1182, May 1994.

27. J. T. Colvin, et al., "Effects of substrate resistances on LNA performance and a bondpad structure for reducing the effects in a silicon bipolar technology," *IEEE J. of Solid-State Circuits*, vol. 34, pp. 1339-1344, Sep. 1999.

28. Y. J. Shin, et al., "An inductorless 900 MHz RF low-noise amplifier in 0.9-μm CMOS," *IEEE Custom Integrated Circuits Conference*, pp. 513-516, May 1997.

29. Q. Huang, et al., "Broadband, 0.25-μm CMOS LNAs with sub-2dB NF for GSM applications," *IEEE Custom Integrated Circuits Conference*, pp. 67-70, May 1998.

30. T.H. Lee, *The Design of CMOS Radio Frequency Integrated Circuits*, Cambridge University Press, 1998.

31. B. Gilbert, "A DC-500MHz amplifier/multiplier principle," *IEEE International Solid-State Circuits Conference Dig. Tech. Papers*, pp. 114-115, Feb 1968.

32. C. Rudell, "Frequency translation techniques for high-integration high-selectivity multi-standard wireless communication systems," Ph.D. thesis, University of California, Berkeley, 2000.

33. S. J. Fang, et al., "A 2GHz CMOS even harmonic mixer for direct-conversion receivers," *IEEE International Symposium on Circuits and Systems*, vol. 4, pp. 807-810, May 2002.

34. S. J. Fang, "Complementary metal-oxide-semiconductor frequency conversion techniques for wideband code division multiple access," Ph.D. Thesis, University of Washington, 2003.

35. S. J. Fang, et al., "An image rejection down-converter for low-IF receivers in 130 nm CMOS," *Radio Frequency Integrated Circuits Symposium*, pp. 57-60, June 2004.

36. D. B. Leeson, "A simple model of feedback oscillator noise spectrum," *Proc. IEEE*, pp. 320-330, Feb. 1966.

37. A. Hajimiri, et al., *The Design of Low Noise Oscillators*, Kluwer Academic Publishers, Feb. 1999.

38. A. Demir, et al., "Phase noise in oscillators: A unifying theory and a numerical method for characterization," *IEEE Trans. on Circuits and Systems-I: Fundamental Theory and Applications*, vol. 47, pp. 655-674, May 2000.

39. J. Rael, et al., "Physical processes of phase noise in differential LC oscillators," *IEEE Custom Integrated Circuits Conf.*, pp. 21-24, May 2000.

40. M. Tiebout, "Low-power low-phase-noise differentially tuned quadrature VCO design in standard CMOS," *IEEE J. Solid-State Circuits*, vol. 36, pp. 1018-1024, July 2001.
41. P. Andreani, et al., "Analysis and design of a 1.8-GHz CMOS LC quadrature VCO," *IEEE J. Solid-State Circuits*, vol. 37, pp. 1737-1747, Dec. 2002.
42. P. Andreani, "A low-phase-noise low-phase-error 1.8Ghz quadrature CMOS VCO," *IEEE International Solid-State Circuits Conference Dig. Tech. Papers*, pp. 290-291, Feb. 2002.
43. M. Chu, et al. "Design considerations for anti-phase injected quadrature voltage-controlled oscillators," *IEEE International Conference on Electronic Circuits and Systems,* Dec. 2004.
44. D. Ham, et al., "Concepts and methods in optimization of integrated LC VCOs," *IEEE J. Solid-State Circuits,* vol. 36, pp. 896-909, June 2001.
45. R. Aparicio, et al., "A noise-shifting differential Colpitts VCO," *IEEE J. Solid-State Circuits,* vol. 37, pp. 1728-1736, Dec. 2002.

International Journal of High Speed Electronics and Systems
Vol. 15, No. 2 (2005) 429–458
© World Scientific Publishing Company

EQUALIZERS FOR HIGH-SPEED SERIAL LINKS

PAVAN KUMAR HANUMOLU

School of Electrical Engineering and Computer Science, Oregon State University
Corvallis, Oregon 97331, U.S.A
hanumolu@ece.orst.edu

GU-YEON WEI

Electrical Engineering and Computer Science, Harvard University.
Cambridge, Massachusetts 02138, U.S.A
guyeon@eecs.harvard.edu

UN-KU MOON

School of Electrical Engineering and Computer Science, Oregon State University.
Corvallis, Oregon 97331, U.S.A
moon@ece.orst.edu

In this tutorial paper we present equalization techniques to mitigate inter-symbol interference (ISI) in high-speed communication links. Both transmit and receive equalizers are analyzed and high-speed circuits implementing them are presented. It is shown that a digital transmit equalizer is the simplest to design, while a continuous-time receive equalizer generally provides better performance. Decision feedback equalizer (DFE) is described and the loop latency problem is addressed. Finally, techniques to set the equalizer parameters adaptively are presented.

Keywords: Serial Link; Eye Diagram; ISI; Equalizer; Jitter; BER; Transceiver; Noise; DFE; Pre-emphasis.

1. Introduction

Recent advances in integrated circuit (IC) fabrication technology coupled with innovative circuit and architectural techniques led to the design of high performance digital systems. The complex systems are built by combining several ICs consisting of millions of transistors operating at multi-gigahertz frequency. These systems require efficient communication between multiple chips for proper functioning of the whole system. However, the off-chip bandwidth scales[1] at a much lower rate compared to the on-chip bandwidth,[2] thus making the communication link (also referred to as serial link) between chips the major bottleneck for the overall performance. For example, present day microprocessors run at several gigahertz clock rates, while the speed of the front-side bus is limited to less than a gigahertz. Due to these reasons, there is a great research interest to reduce the gap between the on-chip and off-chip bandwidth. A representative depiction of a communication link between two chips is shown in Fig. 1. Dedicated circuits designed for high-speed operation

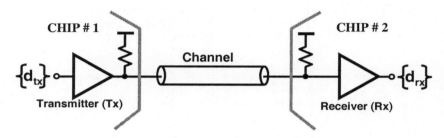

Fig. 1. A typical serial link block diagram.

are used as transmitter and receiver, to transmit and receive the data, respectively. The medium of transmission is called the channel which in the ideal case is a wire representing a short circuit. However, as the data rates increase, these wires behave as lossy transmission lines severely degrading the transmitted data symbols. Equalization is a well-known technique used to overcome non-idealities introduced by the channel. In this paper we present several equalization techniques that are amenable for high-speed operation. The organization of the paper is as follows. Section 2 briefly discusses different aspects of channel modeling, while different metrics used to evaluate the performance of the serial link are summarized in Section 3. Some equalizer background is presented in Section 4, and transmitter and receiver equalizer designs are considered in Section 5 and Section 6, respectively. The advantages of combined transmitter and receiver equalizers are presented in Section 7. Finally, the process of adapting of filter tap weights is discussed in Section 8.

2. Channel Modeling

There are several types of channels used in high-speed interconnects, primarily based on the target application. These channels can be broadly classified into three categories. First, for chip-to-chip communication on a printed circuit board (PCB), short well-controlled copper traces are used. Second, for systems such as local-area network (LAN) which require high-speed connection between two computers, coaxial cable is used as the transmission medium. Finally, copper traces along with backplane connectors are used for high-speed board-to-board communication systems such as routers. In this paper, we focus on the copper traces used for chip-to-chip communication. However, the analysis and the design techniques are easily applicable to a wide range of other channels.

The copper traces commonly used on PCBs behave as lossy transmission lines at multi-gigahertz frequency range. The distributed nature of these transmission lines can be captured by a cascade of an infinitesimal length RLGC sections shown in Fig. 2.[3] Accurate modeling of transmission lines with RLGC sections require

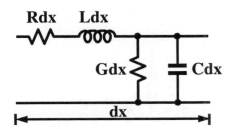

Fig. 2. RLGC section of a transmission line.

quantizing both space and time into sections that are small compared to the shortest wavelength of interest. Therefore, for high-speed designs a large number of RLGC sections are required to comprehend all the transmission line effects. The dominant sources of loss in these channels are due to the skin effect and dielectric loss.[4,5,6] The loss due to skin effect is proportional to $\sqrt{f}$ and typically dominates the total loss at low frequencies. On the other hand, dielectric loss is proportional to f, and therefore, determines the total loss at high frequencies. The lumped RLGC sections are modified as shown in Fig. 3 to comprehend the frequency dependent nature of these loss mechanisms.[7,8] However, with the increased number of nodes in each lumped section, coupled with the requirement of large number of cascaded sections to model the full channel, results in tremendous increase in the overall simulation time. These frequency dependent loss mechanisms also greatly depend on several factors such as the geometry of the traces, making the development of a generic channel model impractical. Due to these issues, channel models are most commonly developed by fitting measured data of each of the components. For example, the s-parameters of the PCB trace with a given geometry are determined by using field solvers such as ADS[9] and are combined with connector models supplied by the vendor to obtain the s-parameters for the complete channel. The S_{21} of a 20″ differential micro-strip line on a FR4 board with two connectors, referred to as *server channel*, and a 6″ differential micro-strip line on the same FR4 board indicated as *desktop channel* is shown in Fig. 4. The s-parameters can be directly used in circuit-level simulators such as SPECTRE or equivalently an impulse response can be derived for system level simulators such as MATLAB. The second approach is more amenable for fast system-level simulations to evaluate the performance of various equalization and clock recovery schemes. Due to these reasons, the impulse response approach is used in this paper. The derived impulse response for the *server* and the *desktop* channel is shown in Fig. 5. Since most practical channels are linear time-invariant systems, an impulse response is sufficient to completely characterize

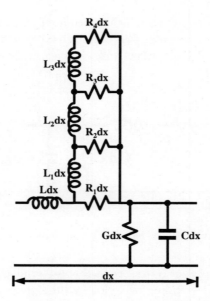

Fig. 3. RLGC section of a transmission line modeling frequency dependent loss.

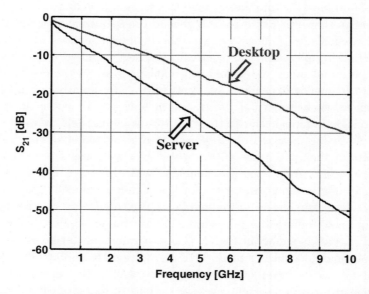

Fig. 4. S_{21} of a 20″ (server) and 6″ (desktop) FR4 trace with two connectors.

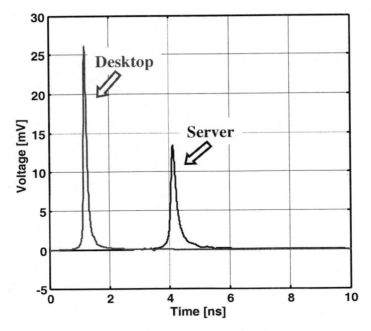

Fig. 5. Impulse response of *server* and *desktop* channels.

it. Before presenting various equalizer designs, we first briefly discuss the metrics used to evaluate the performance of high-speed serial links.

3. Performance Metrics

The primary performance metric in all applications employing serial links is bit error rate (BER). Most systems of interest require almost error-free operation (BER $< 10^{-12}$). However, direct evaluation of such low BER in simulation is not a trivial task. So we will employ an indirect method in which we will calculate the noise margins and relate them to error-free operation. The biggest noise source in high-speed serial links is inter-symbol interference (ISI)[a] caused by the frequency dependent attenuation of the channel.[10] The noise margin degradation due to ISI is best quantified by an eye diagram. The eye diagrams at the receive end of the *server* channel obtained by transmitting 500 pseudo-random bits at 2Gbps and 5Gbps data rates are shown in Fig. 6(a) and Fig. 6(b) respectively. Note that the pseudo-random eye approaches worst-case eye only with very large number of transmitted bits. Even though these eye diagrams clearly indicate the noise margin degradation due to ISI at higher data rates, this approach has two main drawbacks. First, the accu-

---

[a]Even though ISI is fully deterministic, we will refer to it as *noise* here without adhering to the definition of noise in the strict sense.

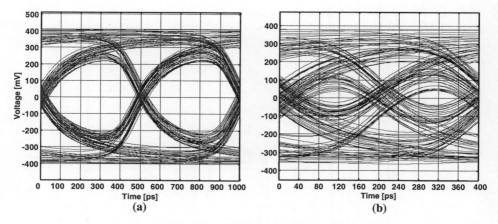

Fig. 6. (a) 2Gbps eye diagram. (b) 5Gbps eye diagram.

rate estimation of worst-case noise margin degradation requires transmitting several thousands of bits, thus increasing the simulation times drastically. Second, this approach does not provide any design insight. We, therefore, employ an analytical method based on pulse response to evaluate the noise margins. Since ISI is completely deterministic in nature, it is possible to calculate the worst case noise margin degradation due to ISI. Consider the representative positive pulse response shown

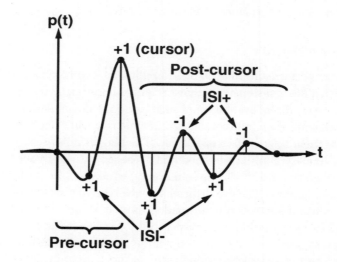

Fig. 7. Conceptual pulse response illustrating ISI terms.

in Fig. 7. The *trailing* ISI and *leading* ISI terms are referred to *pre-cursors* and *post-cursors*, respectively. The worst case effect of these ISI terms on the overall voltage margin at a given sampling instant is obtained simply by adding them in an absolute sense as shown in "Eq. (1)":[11]

$$
\begin{aligned}
\text{Worstcase ISI noise} \; &= \; \sum |ISI_+| + \sum |ISI_-| \\
&= \; \sum ISI_+ - \sum ISI_- \\
&= \; \sum_{k=-\infty}^{\infty} p(t-kT)\big|_{p(t-kT)>0,\,k\neq0} \\
&\quad - \sum_{k=-\infty}^{\infty} p(t-kT)\big|_{p(t-kT)<0,\,k\neq0}
\end{aligned}
\tag{1}
$$

where ISI_- and ISI_+ are negative and positive ISI terms. The complete worst-case ISI eye diagram can be obtained by sweeping the sampling instance across the whole bit period. Fig. 8 displays the worst-case ISI eye diagram calculated by the described method and the eye diagram obtained by transmitting 500 pseudo-random bits. Even though, worst-case eye results in a pessimistic BER prediction, we will

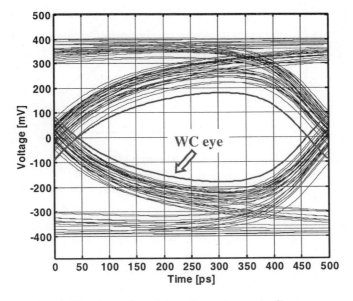

Fig. 8. Pseudo-random data and worst-case eye diagrams.

continue using it in this paper for its simplicity. Interested readers can refer to statistical approaches outlined in[11,12,13,14,15] for more accurate prediction of BER. Other noise sources of concern include circuit noise, clock jitter induced noise,[15,16] and power supply noise.[17] These noise sources are implementation dependent and so we will discuss them separately for each equalizer. In addition to the noise margins, other metrics of interest are circuit area, power consumption and ease of design.

4. Equalizers Background

The magnitude response of the server channel shown again in Fig. 9 illustrates high frequency attenuation due to both skin effect and dielectric loss. For example, the loss for 5 gigabit operation is approximately 12dB resulting in an almost closed eye (see Fig. 6(b)). The frequency shaping filters that flatten the channel response till Nyquist frequency are called equalizers. These equalizers, therefore, reduce ISI and can increase the achievable data rates tremendously. The conceptual diagram illustrating two ways of performing equalization is shown in Fig. 9. In the first

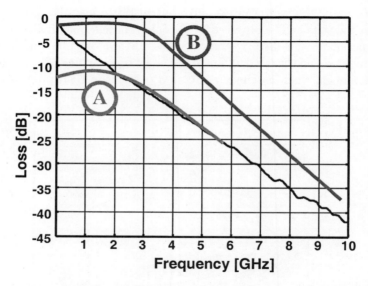

Fig. 9. Two ways of equalization: (a) Attenuate low frequency. (b) Boost high frequency.

method denoted by A, the low-frequencies of the signal spectrum are attenuated, while in the second method denoted by B, the high frequency signal spectrum is boosted in order to mitigate ISI. We now present several techniques and design tradeoffs in implementing these two types of equalizers.

5. Transmit Equalizers

Equalization can be performed either at the transmitter, or at the receiver, or both. In this section we will focus on the transmitter-side equalization. The transmit equalizer shown in Fig. 10 is a symbol spaced (symbol-period is denoted by Δ) finite impulse response (FIR) filter that pre-shapes or pre-distorts transmitted data so as to attenuate the low frequency portion of the signal spectrum while maintaining the high-frequency part intact. Because of this, the transmit equalizers are also referred to as de-emphasis, pre-emphasis, pre-distortion or pre-coding filters.[18,19,20,21,22,23,24] Fig. 11 depicts the eye diagram at 5Gbps equalized with

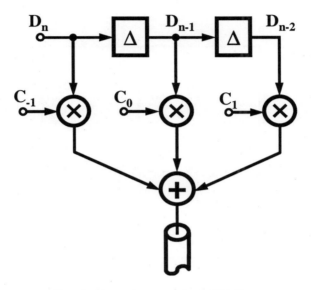

Fig. 10. Transmit pre-emphasis FIR filter.

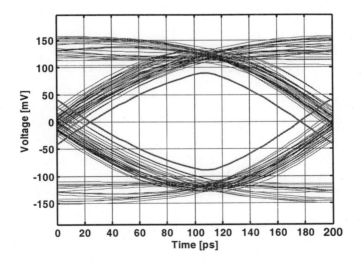

Fig. 11. 5Gbps eye diagram with 3-tap transmit pre-emphasis.

a transmit pre-emphasis filter C = [-0.13 0.66 -0.21]. Post equalization worst case ISI eye shown in Fig. 11 displays 80ps of timing margin with at least 100mV of voltage margin. It is instructive to view the time-domain response of this technique to better visualize the concept. The raw and equalized sampled pulse responses are shown in Fig. 12. Note that the cursor tap is attenuated while at the same time

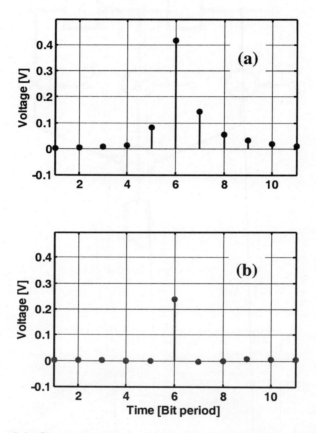

Fig. 12. Sampled 5Gbps pulse response: (a) Raw channel. (b) Transmit pre-emphasis.

the pre-cursor and post cursor ISI is greatly reduced. This cursor-tap attenuation is due to peak transmit power constraint. Consider a practical implementation of a 2-tap pre-emphasis filter shown in Fig. 13, where the tap weights are implemented by scaled tail current sources. These current sources are adjusted digitally by current-mode digital-to-analog converter (DAC), not shown in the figure. It is important to maintain the tail current sources in saturation to prevent reflections introduced by imperfect source termination. Therefore, the maximum output swing of the current mode driver is limited by the voltage headroom needed to maintain

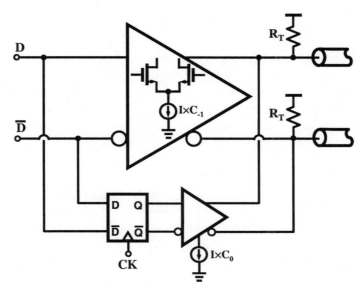

Fig. 13. Two-tap pre-emphasis filter implementation.

high output impedance. Hence, extra taps can be added only at the expense of reducing the cursor tap weight. In other words, since the maximum voltage drop across the termination resistor $I \cdot R_T$ is determined by the voltage headroom, the coefficients should satisfy:

$$\left(I \cdot \sum |C_i|\right) \cdot R_T = I \cdot R_T \Rightarrow \sum |C_i| = 1. \tag{2}$$

There are several limitations of transmit pre-emphasis. First, due to the signal attenuation transmit pre-emphasis can not improve SNR. Second, it is essential to maximize transmitted signal swings to incorporate large amount of equalization, thus resulting in excessive crosstalk.[25] Third, high resolution DACs are required to implement pre-emphasis filters to equalize channels containing large number of ISI terms.[26] Finally, despite transmit pre-emphasis there is considerable residual ISI which results in reduction of both timing and voltage margins, particularly at higher data rates.

6. Receive Equalizers

Receive-side equalization offers an alternate method to mitigate ISI without any peak power constraint. The loss in the channel is suppressed by boosting the high frequency signal spectrum rather than attenuating the low-frequency content. Due to the inherent gain in the system this method often results in larger noise margins. We now present different receive equalizer architectures.

6.1. *Digital FIR equalizer*

Linear transversal filter similar to the one used for transmit pre-emphasis can be used on the receive-side to perform equalization. Unlike the transmit pre-emphasis where the input to the filter is a binary signal, the input to the receive filter is channel output which is analog in nature. An analog to digital converter (ADC) is required to interface the channel output to the filter as shown in Fig. 14. Symbol-spaced delay is implemented by a register. Fig. 15 depicts 5Gbps eye diagram equal-

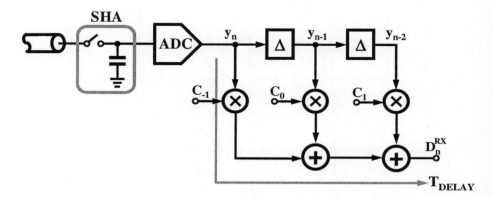

Fig. 14. Digital FIR equalizer.

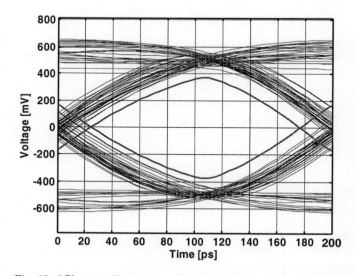

Fig. 15. 5Gbps eye diagram equalized with ideal receiver equalizer.

ized by an ideal 3-tap receive FIR equalizer. The higher voltage margin obtained by receive equalization is clearly evident. However, there are two major bottlenecks in the practical implementation of this equalizer. First, the critical path shown in Fig. 14 limits the maximum operation frequency to only few hundred megahertz. Well known techniques such as transposition[28] and parallelism[32] can be used to shorten critical path. Nevertheless, these transposed filters are still speed-limited to less than a gigabit data rate. Second, the practical usefulness of this equalizer is severely limited by the high-speed ADC requirement at the front-end. Even though high-speed ADCs are possible to design,[27] they add large power and area overhead. Due to these constraints, digital FIR equalizers are employed only in medium rate interfaces such as broadband modems,[29] disk-drive read channels,[30,31,32,33,34] and gigabit ethernet.[35,36] The price paid for high speed operation using digital FIR is excessive power consumption. For example, the equalizer in[34] consumes 1.2W for 2.3Gbps operation in $0.18\mu m$ CMOS process. This amount of power consumption is unacceptable in most serial-link applications which require integrating hundreds of equalizers on a single chip.

6.2. *Analog FIR equalizer*

An analog FIR equalizer obviates the need for a high-speed ADC and is therefore attractive for high-speed operation with potentially lower power consumption. A conceptual block diagram of an analog FIR equalizer is shown in Fig. 16. Note that

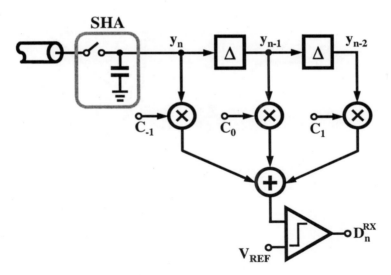

Fig. 16. An analog FIR equalizer.

the high-speed ADC is replaced by a relatively simple sample and hold amplifier (SHA) circuit. As opposed to a digital delay in the case of digital FIR an analog delay chain is required to implement the analog FIR. This analog delay can be im-

plemented using a replica delay line whose delay is locked to a delay locked loop
or a phase locked loop operating at data rate.[37,38] However, the FIR analog equalizer in its most primitive form suffers from many implementation difficulties. First,
the settling time of the front-end SHA limits the overall operating speed. Second,
the sampled signal experiences considerable attenuation due to the limited bandwidth of the delay elements in the delay chain. Moreover, this limited bandwidth
induced error accumulates along the delay chain, thus limiting this technique to
FIR filters with few taps. Finally at high data rates, the precise generation of analog delay consumes excessive power thus negating the primary benefit of an analog
FIR equalizer. One alternate way to generate the analog delay is by using multi-
phase clocks.[39] The delay in the sampling clocks translates to the tap delay. Two
time-interleaved architectures referred to here as Rotating Input Samples (RIS)[41]
and Rotating Tap Weights (RTW)[42] employing multi-phase clocks to implement
tap delay are presented. In the RIS method, a time interleaved N-tap FIR filter
is implemented by using $M > N$ front-end SHAs clocked by multiple phases of a
clock as shown in Fig. 17. The outputs of the M SHAs are routed through a switch

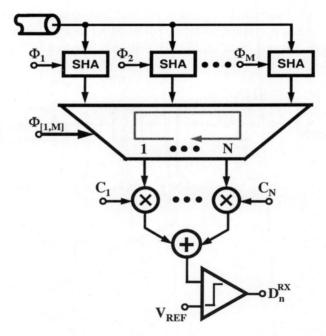

Fig. 17. An analog FIR equalizer based on rotating input samples.

matrix controlled by multi-phase clocks in a circular manner. This circular buffer
architecture increases the minimum settling time of the SHA to more than a clock
period. However, the complexity of the input sample rotating array and mismatches
among various SHAs limit the maximum speed of this architecture to less than

1Gbps data rate.[39,41] Alternately, in the RTW method, the tap weights are rotated instead of the input samples.[42] The conceptual operation of the RTW method is illustrated in Fig. 18 where SR denotes digital shift register. As in the RIS method,

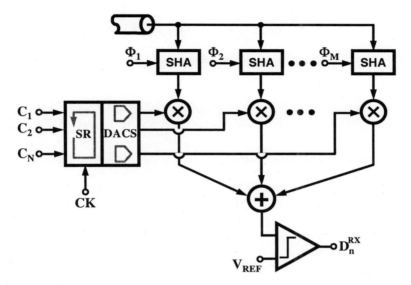

Fig. 18. An analog FIR equalizer based on rotating tap weights.

the circuit employs time interleaved SHAs, but instead shifts the coefficients in a counter-clockwise manner to achieve the FIR filter functionality. By implementing the tap weights using digital words, the RTW method has distinct advantage over the RIS method because the time-interleaving is achieved by rotating digital tap weights, instead of analog input samples. However, RTW method still suffers from the mismatches in the SHAs and does not offer any advantage in high-speed designs employing analog tap coefficients. Parallelism along with time-interleaving can obviate the need for rotating input samples and rotating tap weights and hence permitting very high data rates.[43] The conceptual block diagram of a parallelized architecture is shown in Fig. 19. In this example,[43] the high-speed front-end samplers track the input for two bit periods and hold the sampled value for the next six bit periods. It is well-known that current-mode signal processing can achieve higher speeds, is more efficient and is easier to implement than voltage-mode processing. Therefore, front-end samplers are typically followed by voltage-to-currents (V2I) converters and the operations required for equalization (addition and multiplication) are performed in current domain. Each V2I output is replicated into four interleaved equalizers by simple current mirroring thus accomplishing an effective input sample rotation. The tap-weight multiplication is performed by a current-mode DAC while the summation is achieved by simply shorting the DAC current outputs. Employing parallelism, time-interleaving and current-mode signal process-

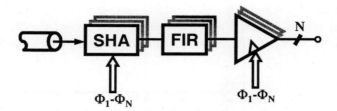

Fig. 19. A parallelized and time-interleaved analog FIR equalizer.

ing, this architecture is suitable for equalizing multi-gigabit serial links, albeit at the expense of increased power and area incurred due to massive parallelism.

6.3. *Continuous time equalizers*

The discrete-time receive equalizers discussed thus far need sampling front-end to perform equalization. This requirement results in two drawbacks. First, the sampling clock-jitter reduces the effectiveness of the equalization. Second, in a truly serial communication system, the clock is recovered from the incoming data. However, due to the sampling front-end, the clock recovery loop needs to operate on raw channel output resulting in an excessive jitter in the recovered clock.[44,45] In order to circumvent the clock recovery problem, practical serial links employing discrete-time FIR equalizers are limited to source synchronous interfaces[43] containing a separate clock channel as shown in Fig. 20. A continuous-time circuit that

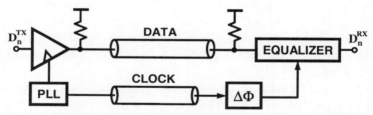

Fig. 20. A source synchronous interface employing discrete-time equalizer. $\Delta\Phi$ compensates for the delay mismatch between clock and data channels.

can provide high-frequency boost is a very attractive alternative to the transversal filters employing sampling front-ends. A continuous-time equalizer is a simple one tap continuous-time circuit with high-frequency gain boosting transfer function that effectively flattens the channel response. As an example, the required frequency shaping can be achieved by a simple RC network as shown in Fig. 21. The resistor attenuates the low-frequency signals while the capacitor allows the high-frequency signal content, thus resulting in high frequency gain boosting. The transfer function

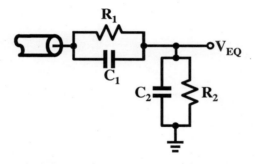

Fig. 21. Continuous-time passive equalizer.

and the pole zero frequencies are given by:

$$H(s) = \frac{R_2}{R_1 + R_2} \frac{1 + R_1 C_1 s}{1 + \frac{R_1 R_2}{R_1 + R_2}(C_1 + C_2)s} \tag{3}$$

$$\omega_z = \frac{1}{R_1 C_1} \tag{4}$$

$$\omega_p = \frac{1}{\frac{R_1 R_2}{R_1 + R_2}(C_1 + C_2)} \tag{5}$$

$$DC\ gain = \frac{R_2}{R_1 + R_2} \tag{6}$$

The gain-boost factor is proportional to the ratio of zero and pole frequency $\frac{\omega_z}{\omega_p}$, so reasonable amounts of equalization can be achieved by choosing appropriate component values that set the required gain-boosting. For example, the equalizer obtained with $R_1 = 200\Omega$, $C_1 = 1pF$, $R_2 = 65\Omega$ and $C_2 = 0.1pF$, results in considerable eye-opening at 5Gbps on the *server* channel as shown in Fig. 22. There are two main disadvantages with simple passive RC equalizers. First, the RC network introduces large impedance discontinuity at the channel and equalizer interface. Impedance matching networks,[54] often employing inductors, can be used to prevent the discontinuity. However, the large inductors make this approach less suitable for on-chip integration. Second, this method can not improve SNR, since equalization is performed by attenuating low-frequency signal spectrum much like transmit pre-emphasis. Due to these reasons, this technique has limited use in high-speed serial links.

It is desirable to have a gain greater than one at all frequencies to maximize the benefit from receiver-side equalization. Therefore, equalizers using active circuit elements rather than passive components are required to achieve gains greater than one. Active filters with desired frequency response can be designed using standard filter design techniques.[46] Such standard filters are typically implemented either with operational amplifiers in negative feedback or Gm-C filter topology. However, the negative feedback as used in these systems greatly degrades the maximum oper-

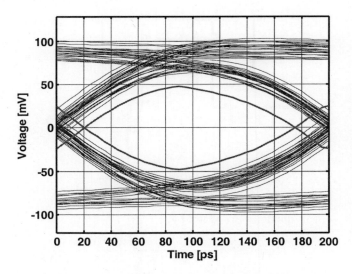

Fig. 22. 5Gbps eye diagram using passive equalizer.

ating frequency thus, limiting the usefulness of such equalizers to only few hundred megahertz.[47,48,49] Recent attempts however focus on open loop equalizer architectures to overcome bandwidth penalty due to negative feedback.[50] There are several wide-band amplifier design techniques that can provide the required high frequency boost for equalization. These techniques include bandwidth enhancement by zeros, and tuned and/or cascaded amplifiers.[54]

As noted earlier in the design of passive equalizer, the parallel RC combination introduces real zero in the transfer function, potentially providing gain-peaking. The active-equivalent of the passive equalizer can be designed by degenerating a source-coupled pair with the parallel RC network as shown in Fig. 23.[51,52] The transfer function and the associated pole-zero locations are given by:

$$H(s) = \frac{g_m}{C_L} \frac{s + \frac{1}{R_D C_D}}{\left(s + \frac{g_m R_D + 1}{R_D C_D}\right)} \cdot \frac{1}{\left(s + \frac{1}{R_L C_L}\right)} \tag{7}$$

$$\omega_z = \frac{1}{R_D C_D} \tag{8}$$

$$\omega_{p1} = \frac{g_m R_D + 1}{R_D C_D} \tag{9}$$

$$\omega_{p2} = \frac{1}{R_L C_L s} \tag{10}$$

$$DC\ gain = \frac{g_m R_L}{g_m R_D + 1} \tag{11}$$

By designing the zero frequency to be lower than the dominant pole, considerable

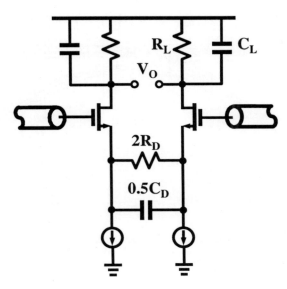

Fig. 23. Continuous-time equalizer using capacitive degeneration. R_L and C_L represent the load.

high frequency gain boosting can be achieved. The amount of this gain boost can be controlled by the ratio of the dominant pole and zero frequency (ω_z/ω_{p1}). fcaptionChannel response: (a) Raw. (b) With continuous-time equalizer.

The continuous-time equalizer implemented in a $0.18\mu m$ CMOS process operating with 1.8V supply and consuming less than 10mW of power, provides a gain boost of 8dB at 2.5GHz as depicted in Fig. . The equalized eye diagram at 5Gbps data rate shown in Fig. 24 displays 120ps of timing margin with at least 100mV of voltage margin. Thus, compared with transmit pre-emphasis, continuous-time receive equalizer provides 65% more timing margin reinforcing the benefit of receive-side equalization. The maximum gain boosting achieved by this method is limited by the bandwidth of the amplifier due to the load capacitance (ω_{p2}). Inductive peaking shown in Fig. 25(a)[53] or neutralization as shown in Fig. 25(b),[54] can be used to increase the amplifier bandwidth and hence improve gain-boost factor. Also, a cascade of these equalizer stages can provide higher gain without sacrificing the bandwidth.[54,55] It is worth mentioning that gain peaking can also be achieved by zeros introduced by the load inductor as shown in Fig. 26.[56] The equalizer output is the weighted sum of the flat gain amplifier output and the gain peaked amplifier output. Implemented in 150GHz f_T BiCMOS process, this equalizer provides almost 30dB gain boost at 7GHz and achieves 10Gbps data rate on a channel consisting of 15 feet of coaxial cable. The large gain-boosting using a single stage was possible due to large (9GHz) amplifier bandwidth. However, this equalizer consumes 200mW of power, making it less attractive for main-stream serial links.

Finally, continuous-time transversal filters, as opposed to discrete-time filters,

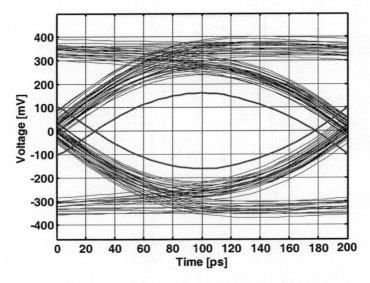

Fig. 24. Continuous-time receive equalized 5Gbps eye diagram.

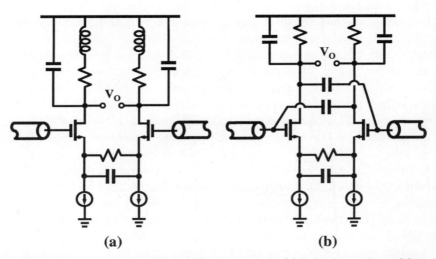

(a)   **(b)**

Fig. 25. Continuous-time equalizer bandwidth enhancement: (a) Inductive peaking. (b) Neutral-ization.

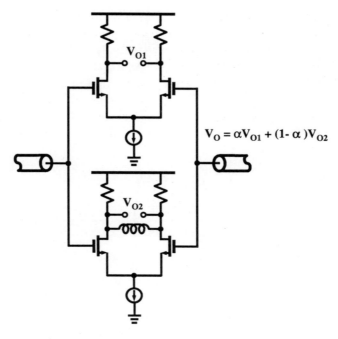

Fig. 26. Continuous-time equalizer using inductor load.

can be implemented if one can design high-bandwidth analog-delay elements. A 10Gbps continuous-time analog FIR equalizer using distributed techniques to generate the analog delay is recently proposed.[57] The precise delay is generated by using transmission-line sections shown in Fig. 27. Even though this method has a

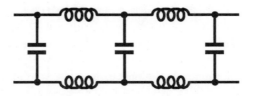

Fig. 27. Transmission-line delay element.

potential high speed advantage, it is not practical at medium to high data rates (5 to 10Gbps) due to the requirement of very long well-controlled on-chip transmission lines or large number of area consuming inductors.[57]

6.4. *Noise enhancement*

We have thus far presented techniques for suppressing ISI, without alluding to other noise sources in the system. In this section we focus on the noise introduced by the continuous-time equalizer itself. The gain-peaking transfer function of the equalizer amplifies the high frequency noise potentially degrading the noise margin. Also in equalizers employing multiple stages, the first stage generally dominates the overall noise.[b] We now estimate the noise contribution of a single equalizer stage shown in Fig. 23 The output noise is typically dominated by the input transistor pair. The one-sided voltage noise power spectral density of the input transistor given by $\overline{V_{in}^2} = 4kT\gamma / g_m$ is amplified by the equalizer transfer function resulting in an total output noise

$$
\begin{aligned}
\overline{V_{on,Total}^2} &= \int_0^\infty |H(s)|^2 \, \overline{V_{in}^2} df \\
&= \frac{g_m}{C_L} \frac{\sqrt{\omega_{p1} \cdot \omega_{p2}}}{2(\omega_{p1} + \omega_{p2})} \left[1 + \frac{\omega_z}{\sqrt{\omega_{p1} \cdot \omega_{p2}}}\right] \overline{V_{in}^2}
\end{aligned}
\tag{12}
$$

For the equalizer used to achieve about 8dB gain boost at 2.5GHz in $0.18\mu m$ CMOS technology, the output referred rms noise voltage $\sqrt{\overline{V_{on,Total}^2}}$ is less than $1mV_{rms}$. For links operating with at least several tens of millivolts of signal swings, this noise enhancement does not limit the overall performance.

6.5. *Decision feedback equalization*

The problem of noise enhancement can be completely eliminated by using Decision Feedback Equalizer (DFE) shown in Fig. 28. Unlike the aforementioned equalizers, DFE utilizes the previous decisions to estimate and cancel the ISI introduced by the lossy channel. The feedback filter, estimates the ISI based on previous decisions, and therefore, can only cancel post-cursor ISI (i.e. ISI caused by previous symbols). Since the ISI cancellation is based on previous decisions, without high-frequency boost, it is inherently immune to noise enhancement. There are three design issues with the DFE design. First, the effectiveness of ISI cancellation is based on the assumption that all previous decisions are correct and therefore bit errors can exacerbate ISI instead of cancelling it. This problem is referred to as error propagation. However, in the case of serial-links with required BER $< 10^{-12}$ error propagation does not degrade the performance.[58] Second, DFE can cancel only post-cursor ISI, and therefore, a separate feed-forward filter is required to cancel pre-cursor ISI. The analog FIR equalizer or transmit pre-emphasis [25,64] optimized to cancel pre-cursor ISI can be used. Finally, the DFE implementation suffers from the feed-back loop latency illustrated as critical timing path in Fig. 28. The loop latency due to the input slicer regeneration time and the coefficient DAC settling time should be less than the bit period in order for the feedback to cancel the first post cursor ISI. However, at high data rates, this loop delay is more than several bit periods. Decision look-ahead schemes as shown for a single bit case in Fig. 29 are

[b]In an equalizer employing cascaded stages, low frequency input referred noise can be amplified if the DC gain is less than one.

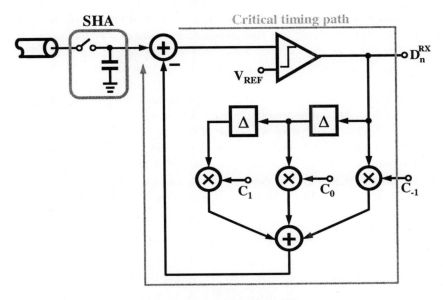

Fig. 28. Decision feedback equalizer.

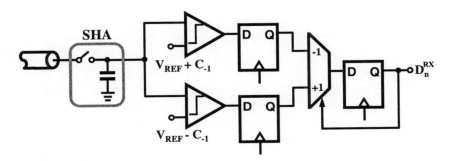

Fig. 29. 1-bit decision look-ahead feedback equalizer.

used to circumvent the latency issue.[59,60,61,62,63] Two parallel receivers resolve the channel output for the two possible previous outputs (+1,-1). The correct output is then selected by the previous bit using a simple multiplexor. Note that the hardware to implement this look-ahead scheme grows exponentially with the number of taps, thus limiting it to only few taps in practical implementations.

7. Transmit and Receive Equalizers

We have considered transmit and receive equalizers independently thus far. As mentioned earlier, transmit pre-emphasis suffers from peak power constraint and the receive equalizer performance is constrained by several non-idealities such as the limited amplifier bandwidth, noise enhancement and amplifier non-linearity. However, some of these issues can be circumvented by using both transmit and receive equalizers together. Employing a 3-tap transmit pre-emphasis filter and a single stage receive equalizer the eye diagram shown in Fig. 30 displays 50ps of

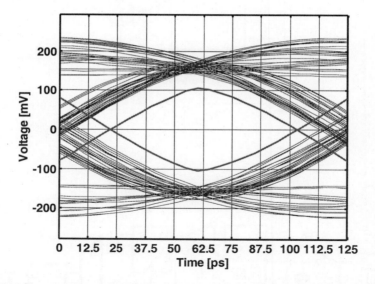

Fig. 30. 8Gbps eye diagram with both transmit pre-emphasis and receive equalization.

timing margin with at least 100mV voltage margin. The amount of equalization performed by receiver and the transmitter can be evaluated by the pulse responses shown in Fig. 31. Fig. 31(a) displays the raw 8Gbps pulse response clearly illustrating both pre-cursor and post-cursor ISI. Fig. 31(b) depicts the receive equalized pulse response. There are two important things to note. First, the receive equalizer amplifies the cursor tap while attenuating post-cursor ISI. Second, the dominant left-over ISI is due to first pre-cursor and post-cursor terms. Therefore, a 3-tap pre-emphasis is used to reduce these ISI terms as shown in Fig. 31(c).

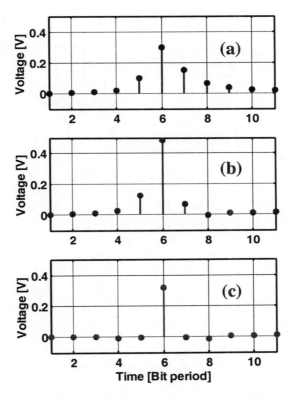

Fig. 31. 8Gbps pulse response: (a) Raw. (b) Receive equalized. (c) Receive and transmit equalized.

8. Adaptation

In a practical transmission system, the exact channel characteristics are not known a priori. Therefore, making the pre-designed equalizer grossly sub-optimal. For example, the channel length can vary from one application to another, or the loss profile of the channel may vary due to the variations in the PCB fabrication process. Due to these reasons, the equalizer coefficients are set adaptively. The conceptual block diagram illustrating the operation of an adaptive equalizer is shown in Fig. 32. In this generic block diagram the equalizer could be of any type —

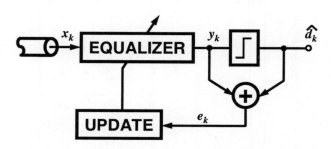

Fig. 32. Adaptive equalizer concept.

discrete-time FIR, continuous-time FIR, or continuous-time analog equalizer. The adaptive engine automatically adjusts the coefficients by measuring the equalizer performance so as to improve the performance on an average.

There are several algorithms[65] that can be used for adapting the equalizer. Of these the most popular ones for compact hardware implementation are the Least Mean Squares (LMS) and the Zero-Forcing (ZF) algorithms or their variants. The LMS algorithm optimizes the filter coefficients based on minimizing the mean squared error. The coefficient update equation in the LMS algorithm is given by "Eq. (13)":

$$c_{(k+1,n)} = c_{(k,n)} + \mu \cdot e_k \cdot x_{k-n} \ \ for \ n = 0 \cdots N \tag{13}$$

where $c_{(k+1,n)}$ represents n^{th} filter coefficient with N taps at $(k+1)^{1st}$ update, μ is the update step size and $e_k = y_k - \widehat{d_k}$ is the error in the equalizer output, $\widehat{d_k}$ the best estimate of the transmitted bit and x_k is the channel output. The analog multipliers are required to realize the update equation (e_k and x_k are analog in nature), making the hardware implementation of the update equation difficult. The sign-sign LMS update algorithm is given by:

$$c_{(k+1,n)} = c_{(k,n)} + \mu \cdot sign(e_k) \cdot sign(x_{k-n}) \ \ for \ n = 0 \cdots N \tag{14}$$

"Eq. (14)" obviates the need for an analog multiplier, thus making it more amenable for on-chip integration.[58,66,67] The adaptation engine consists of UP/DOWN counters that count up or down according to the product of the error sign and the data sign. Since a quantized error is used to update the tap weights, the convergence time of sign-sign LMS is generally worse than the traditional LMS algorithm. In most serial-link applications, this increased convergence time is not a problem. Training sequences, as shown in Fig. 33 are used to ease the convergence of the

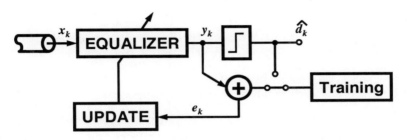

Fig. 33. Adaptive equalizer with a training sequence.

tap weights. After training for a time period sufficient for tap weight convergence, the training sequence is replaced by the decision of the receiver for adaptation to continue.[67] Continuous-time analog equalizers can also be adapted using similar concepts but require different error detection mechanisms. Interested readers can refer to[47,48,49,52,68] for more information.

9. Conclusions

We have presented different equalizer architectures suitable for medium-to-high data rate serial-link applications. Design trade-offs in both transmit-side and receive-side equalizers were presented. Transmit pre-emphasis suffers from peak power constraint while receive equalizer performance is limited by amplifier bandwidth. The availability of high f_T transistors coupled with the need for large amounts of equalization will make receive equalization a very attractive alternative in the future. Finally, techniques to adaptively set equalizer settings were described.

References

1. D. Schmidt, "Circuit pack parameter estimation using Rent's rule," *IEEE Trans. on Computer Aided Design of Integrated Circuits and Systems*, pp. 186-192, Oct. 1982.
2. G. Moore, "Cramming more components onto integrated circuits," *Electronics*, pp. 114-117, April, 1965.
3. D. Pozar, *Microwave Engineering*, Second edition, John Wiley & Sons, Inc., 1998.
4. W. Dally and J. Poulton, *Digital Systems Engineering*, Cambridge University Press, 1998.
5. H. Bakoglu, *Circuits, Interconnections, and Packaging for VLSI*, Addison-Wesley Publishing Company, 1990.
6. H. Johnson and M. Graham, *High-Speed Digital Design*, Prentice Hall, 1993.
7. S. Kim and D. P. Neikirk, "Compact equivalent circuit model for the skin Effect," *IEEE MTTS Dig. Tech. Papers*, pp. 17-21, Jun. 1996.

8. H. Wheeler, "Formulas for the skin-effect," *Proc. Institute of Radio Engineers*, pp. 412-424, 1942.

9. "Advanced Design Systems," *ADS Design Guides*, Agilent Technologies, 2000.

10. E. Lee and D. Messerschmitt, *Digital Communication*, Kluwer Academic Publishers, 1994.

11. B. Casper, M. Haycock, and R. Mooney, "An accurate and efficient analysis method for multi-Gb/s chip-to-chip signaling schemes," *IEEE VLSI Circuits Sym. Tech. Papers*, pp. 54-57, Jun. 2002.

12. B. Ahmad, "Performance specification of interconnects," *DesignCon*, 2003.

13. R. Kollipara, G. Yeh, B. Chia, A. Agarwal, "Design, modeling and characterization of high-speed backplane interconnects," *DesignCon*, 2003.

14. S. Sercu and J. De Geest, "BER Link Simulations," *DesignCon*, 2003.

15. V. Stojanovic, M. Horowitz "Modeling and analysis of high-speed links," *Proc. of IEEE CICC*, pp. 589-594, Sep. 2003.

16. P. Hanumolu, B. Casper, R. Mooney, G. Wei and U. Moon, "Analysis of PLL clock jitter in high-speed serial links," *IEEE Trans. Circuits Syst. II*, vol. 50, no. 11, pp. 879-886, Nov. 2003.

17. E. Alon, V. Stojanovic, M. Horowitz, "Circuits and techniques for high-resolution measurement of on-chip power supply noise," *IEEE VLSI Circuits Sym. Tech. Papers*, pp. 102-105, Jun. 2004.

18. W. Dally and J. Poulton, "Transmitter Equalization for 4-Gbps Signaling," *IEEE Micro*, pp. 48-56, 1997.

19. A. Fiedler, R. Mactaggart, J. Welch, and S. Krishnan, "A 1.0625Gbps transceiver with 2x-oversampling and transmit signal pre-emphasis," *ISSCC Dig. Tech. Papers*, pp. 238-239, Feb. 1997.

20. R. Gu, J. Tran, H. Lin, A. Yee, and M. Izzard, "A 0.5-3.5Gb/s low-power low-jitter serial data CMOS transceiver," *ISSCC Dig. Tech. Papers*, pp. 352-353, Feb. 1999.

21. B. Casper, A. Martin, J. Jaussi, J. Kennedy, and R. Mooney, "8Gb/s differential simultaneous bidirectional link with 4mV 9ps waveform capture diagnostic capability," *ISSCC Dig. Tech. Papers*, pp. 78-79, Feb. 2003.

22. R. Farjad-Rad, C. Yang, and M. Horowitz, "A $0.4 - \mu m$ CMOS 10-Gb/s 4-PAM pre-emphasis serial link transmitter," *IEEE J. Solid-State Circuits*, vol. 34, no. 5, pp. 580-585, May 1999.

23. M. Lee, W. Dally, P. Chiang, "A 90 mW 4 Gb/s equalized I/O circuit with input offset cancellation," *ISSCC Digest Technical Papers*, Feb. 2000.

24. B. Lee, M. Hwang, S. Lee, and D. Jeong, "A 2.5-10Gb/s CMOS transceiver with alternating edge sampling phase detection for loop characteristic stabilization," *ISSCC Dig. Tech. Papers*, pp. 76-77, Feb. 2003.

25. J. Zerbe, P. Chau, C. Werner, W. Stonecypher, H. Liaw, G. Yeh, T. Thrush, S. Best, K. Donnelly, "A 2Gb/s/pin 4-PAM parallel bus interface with transmit crosstalk cancellation, equalization, and integrating receivers," *ISSCC Dig. Tech. Papers*, pp. 66-67, Feb. 2001.

26. J. Zerbe, C. Werner, V. Stojanovic, F. Chen, J. Wei, G. Tsang, D. Kim, W. Stonecypher, A. Ho, T. Thrush, R. Kollipara, M. Horowitz, K. Donnelly, "Equalization and clock recovery for a 2.5-10-Gb/s 2-PAM/4-PAM backplane transceiver cell," *IEEE J. Solid-State Circuits*, pp. 2121-2130, Dec. 2003.

27. C. Yang, V. Stojanovic, S. Modjtahedi, M. Horowitz, W. Ellersick "A serial-link transceiver based on 8-Gsamples/s A/D and D/A converters in $0.25\mu m$ CMOS', *IEEE J. Solid-State Circuits*, pp. 1684-1692, Nov. 2001.

28. K. Azadet, C. Nicole, "Low-power equalizer architectures for high speed modems,"

IEEE Communication Magazine, pp. 118-126, Oct. 1998.

29. C. Nicol, P. Larsson, K. Azadet, J. O'Neill, "A low-power 128-tap digital adaptive equalizer for broadband modems," *IEEE J. Solid-State Circuits*, pp. 1777-1789, Nov. 1997.

30. D. Moloney, J. O'Brien, E. O'Rourke, F. Brianti, "Low-power 200-Msps, area-efficient, five-tap programmable FIR filter," *IEEE J. Solid-State Circuits*, vol. 33, pp. 1134-1138, July 1998.

31. C. Wong, J. Rudell, G. Uehara, P. Gray, "A 50 MHz eight-tap adaptive equalizer for partial-response channel," *IEEE J. Solid-State Circuits*, vol. 30, pp. 228-234, Mar. 1995.

32. L. Thon, P. Sutardja, F. Lai, G. Coleman, "A 240 MHz 8-tap programmable FIR filter for disk-drive read channels," *ISSCC Dig. Tech. Papers*, pp. 82-83, Feb. 1995.

33. R. Staszewski, K. Muhammad, P. Balsara, " 550 -MSample/s 8 -Tap FIR digital filter for magnetic recording read channels," *IEEE J. Solid-State Circuits*, vol. 35, pp. 1205-1210, Aug. 2000.

34. S. Rylov *et al.*, "A 2.3 GSample/s 10-tap digital FIR filter for magnetic recording read channels," *ISSCC Dig. Tech. Papers*, pp. 190-191, Feb. 2001.

35. E. Haratsch, K. Azadet, " A 1-Gb/s joint equalizer and trellis decoder for 1000BASE-T gigabit Ethernet," *IEEE J. Solid-State Circuits*, vol. 36, pp. 374-384, July 2000.

36. K. Azadet, M. Yu, P. Larsson, D. Inglis, "A gigabit transceiver chip set for UTP CAT-6 cables in digital CMOS technology," *ISSCC Dig. Tech. Papers*, pp. 306-307, Feb. 2000.

37. J. Buckwalter, A. Hajimiri, "An active analog delay and the delay reference loop," *Proc. of IEEE RFIC Symposium*, pp. 17-20, Jun. 2004.

38. J. Yang, J. Kim, S. Byun, C. Conroy, B. Kim, "A quad-channel 3.125Gb/s/ch serial-link transceiver with mixed-mode adaptive equalizer in $0.18\mu m$ CMOS," *ISSCC Dig. Tech. Papers*, pp. 176-177, Feb. 2004.

39. D. Xu, Y. Song, G. Uehara, "A 200MHz 9-tap analog equalizer for magnetic disk read channels in $0.6\mu m$ CMOS," *ISSCC Dig. Tech. Papers*, pp. 74-75, Feb. 1996.

40. R. Farjad-Rad, C. Yang, and M. Horowitz, "A 0.3-μ CMOS 8-Gb/s 4-PAM serial link transceiver," *IEEE J. Solid-State Circuits*, vol. 35, pp. 757-764, May 2000.

41. X. Wang, R. Spencer, "A low-power 170-MHz discrete-time analog FIR filter," *IEEE J. Solid-State Circuits*, vol. 35, pp. 417-426, March 1998.

42. T. Lee, B. Razavi, "A 125-MHz CMOS mixed-signal equalizer for Gigabit Ethernet on copper wire," *Proc. of IEEE CICC*, pp. 131-134, May 2001.

43. J. Jaussi, G. Balamurugan, D. Johnson, B. Casper, A. Martin, J. Kennedy, N. Shanbhag, R. Mooney, "An 8Gb/s Source-Synchronous I/O Link with adaptive receiver equalization, offset cancellation and clock deskew," *ISSCC Dig. Tech. Papers*, pp. 242-243, Feb. 2004.

44. J. Buckwalter, A. Hajimiri, "A 10Gb/s data-dependent jitter equalizer," *Proc. of IEEE CICC*, pp. 39-42, Oct. 2004.

45. J. Buckwalter, B. Analui, A. Hajimiri, "Data-dependent jitter and crosstalk-induced bounded uncorrelated jitter in copper interconnects," *IEEE MTTS Dig. Tech. Papers*, pp. 1627-1630, Jun. 2004.

46. R. Schaumann, M. Van Valkenburg *Design of Analog Filters*, Oxford University Press, 2001.

47. J. Babanezhad, "A 3.3-V Analog adaptive line-equalizer for fast ethernet data connection," *Proc. of IEEE CICC*, pp. 343-346, May 1998.

48. G. Hartman, K. Martin, A. McLaren, "Continuous-time adaptive-analog coaxial cable equalizer in $0.5\mu m$ CMOS," *Proc. of IEEE ISCAS*, pp. 97-100, May 1999.

49. O. Shoaei *et al.*, "A 3V Low-Power 0.25μm CMOS 100Mb/s receiver for fast ethernet," *ISSCC Dig. Tech. Papers*, pp. 308-309, Feb. 1996.

50. Y. Kudoh, M. Fukaishi, M. Mizuno, "A 0.13μm CMOS 5-Gb/s 10-meter 28AWG Cable transceiver with no-feedback-loop continuous-time post-equalizer," *IEEE VLSI Circuits Sym. Tech. Papers*, pp. 64-67, Jun. 2002.

51. R. Farjad-Rad *et al.*, "0.622-8.0 Gbps 150 mW serial IO macrocell with fully flexible pre-emphasis and equalization," *IEEE VLSI Circuits Sym. Tech. Papers*, pp. 63-66, Jun. 2003.

52. J. Choi, M. Hwang, D. Jeong, "A 0.18μm CMOS 3.5-gb/s continuous-time adaptive cable equalizer using enhanced low-frequency gain control method," *IEEE J. Solid-State Circuits*, vol. 39, pp. 419-425, March 2004.

53. S. Mohan, M. Hershenson, S. Boyd, T. Lee, "Bandwidth extension in CMOS with optimized on-chip inductors," *IEEE J. Solid-State Circuits*, vol. 35, pp. 346-355, March 2000.

54. T. Lee, *The Design of CMOS Radio-Frequency Integrated Circuits*, Cambridge University Press, 1998.

55. S. Galal, B. Razavi, "10Gb/s Limiting Amplifier and Laser/Modulator Driver in 0.18μm CMOS technology," *ISSCC Dig. Tech. Papers*, pp. 188-189, Feb. 2003.

56. G. Zhang, P. Chaudhari, M. Green "A BiCMOS 10Gb/s adaptive cable equalizer," *ISSCC Dig. Tech. Papers*, pp. 482-483, Feb. 2004.

57. H. Wu, J. Tierno, P. Pepeljugoski, J. Schaub, S. Gowda, J. Kash, A. Hajimiri, "Integrated transversal equalizers in high-speed fiber-optic systems," *IEEE J. Solid-State Circuits*, vol. 38, pp. 2131-2137, Dec. 2003.

58. V. Balan, J. Caroselli, J. Chem, C. Desai, C. Liu, "A 4.8-6.4 Gbps serial link for backplane applications using decision feedback equalization," *Proc. of IEEE CICC*, pp. 31-34, Oct. 2003.

59. K. Parhi, "High-speed architectures for algorithms with quantizer loops," *Proc. of IEEE ISCAS*, pp. 2357-2360, May 1990.

60. S. Kasturia, J. Winters, "Techniques for high-speed implementation of nonlinear cancellation," *IEEE J. Selected Areas in Communications*, vol. 38, pp. 711-717, Jun. 1991.

61. Y. Sohn, S. Bae, H. Park, C. Kim, S. Cho, "A 2.2 Gbps CMOS look-ahead DFE receiver for multidrop channel with pin-to-pin time skew compensation," *Proc. of IEEE CICC*, pp. 473-476, Sep. 2003.

62. R. Kajley, P. Hurst, J. Brown, "A mixed-signal decision-feedback equalizer that uses a look-ahead architecture," *IEEE J. Solid-State Circuits*, vol. 32, pp. 450-459, Mar. 1997.

63. V. Stojanovic *et al.*, "Adaptive equalization and data recovery in a dual-mode (PAM2/4) serial link transceiver," *IEEE VLSI Circuits Sym. Tech. Papers*, pp. 348-351, Jun. 2004.

64. M. Tomlinson, "New automatic equalizer employing modulo arithmetic," *Electr. Let.*, pp. 138-139, March 1971.

65. J. Proakis, *Digital Communications*, McGraw-Hill Education, 2000.

66. J. Stonick, G. Wei, J. Sonntag, D. Weinlader, "An adaptive PAM-4 5-Gb/s backplane transceiver in 0.25μm CMOS," *IEEE J. Solid-State Circuits*, vol. 38, pp. 436-443, Mar. 2003.

67. G. Balamurugan *et al.*, "Receiver adaptation and system characterization of an 8Gbps source-synchronous I/O link using on-die circuits in 0.13μm CMOS," *IEEE VLSI Circuits Sym. Tech. Papers*, pp. 356-359, Jun. 2004.

68. A. Baker, "An adaptive cable equalizer for serial digital video rates to 400 Mb/s," *ISSCC Dig. Tech. Papers*, pp. 174-175, Feb. 1996.

International Journal of High Speed Electronics and Systems
Vol. 15, No. 2 (2005) 459–476
© World Scientific Publishing Company

LOW-POWER, PARALLEL INTERFACE WITH CONTINUOUS-TIME ADAPTIVE PASSIVE EQUALIZER AND CROSSTALK CANCELLATION

C. PATRICK YUE, JAEJIN PARK, RUIFENG SUN, L. RICK CARLEY

Department of Electrical & Computer Engineering
Carnegie Mellon University
Pittsburgh, PA 15213, USA
{cpyue, jaejinp, ruifengs, rick.carley}@ece.cmu.edu

FRANK O'MAHONY

Circuit Research Lab, Intel Corp.
Hillsboro, OR 97123, USA
frank.o'mahony@intel.com

This paper presents the low-power circuit techniques suitable for high-speed digital parallel interfaces each operating at over 10 Gbps. One potential application for such high-performance I/Os is the interface between the channel IC and the magnetic read head in future compact hard disk systems. First, a crosstalk cancellation technique using a novel data encoding scheme is introduced to suppress electromagnetic interference (EMI) generated by the adjacent parallel I/Os. This technique is implemented utilizing a novel 8-4-PAM signaling with a data look-ahead algorithm. The key circuit components in the high-speed interface transceiver including the receive sampler, the phase interpolator, and the transmitter output driver are described in detail. Designed in a 0.13-μm digital CMOS process, the transceiver consumes 310 mW per 10-Gps channel from a 1-V supply based on simulation results. Next, a 20-Gbps continuous-time adaptive passive equalizer utilizing on-chip lumped *RLC* components is described. Passive equalizers offer the advantages of higher bandwidth and lower power consumption compared with conventional designs using active filter. A low-power, continuous-time servo loop is designed to automatically adjust the equalizer frequency response for the optimal gain compensation. The equalizer not only adapts to different channel characteristics, but also accommodates temperature and process variations. Implemented in a 0.25-μm, 1P6M BiCMOS process, the equalizer can compensate up to 20 dB of loss at 10 GHz while only consumes 32 mW from a 2.5-V supply.

Keywords: crosstalk cancellation; multi-level pulse amplitude modulation; low-power equalizer; continuous-time equalizer, adaptive equalizer.

1. Introduction

The demand for data storage systems with a higher data capacity, smaller form factor, faster access time, and lower system cost has fueled the exponential evolution in hard disk drive (HDD) technology over the past three decades. Today, state of the art systems have achieved storage capacity of 100 Gbit/in^2 with an internal data rate in excess of 1 Gbps [1]. By continuing the reduction of disk diameter, the projected HDD capacity and speed are expected to reach 1 Tbit/in^2 and 10 Gbps, respectively, by 2010. For emerging mobile applications, the miniaturization trend further provides the important benefit of better power utilization by lowering energy loss due to air drag in the HDD system.

To keep pace with the system advancements, future generations of read-channel ICs must deliver greater capability in digital signal processing (DSP) and support higher bandwidth data interface at low-power consumption. In conventional HDD systems as illustrated in Fig. 1(a), the data communication between the front-end IC (pre-amplifier and write driver) and channel IC is analog. As the data rate increases, analog signaling becomes more vulnerable to noise. Channel ICs have benefited from CMOS device scaling since the DSP circuit performance improves. However, the reduced supply voltage and degraded analog device characteristics make it increasingly difficult to implement the analog interface circuits. Recently, a channel IC with 2.1-Gbps analog interface was reported in a 0.13-μm CMOS process [2]. However, dedicated analog transistors, in addition to the standard logic and I/O devices, were used for the interface circuits. The extra mask and silicon cost incurred by this approach may not be justified since the digital portion of the channel IC totally dominates die area.

To overcome these problems, we propose an alternative architecture by migrating the analog interface circuits from the channel IC into the front-end IC as shown in Fig. 1(b). This approach relaxes the analog interface circuit design as the front-end IC is typically implemented in a more analog-friendly process such as a Bi-CMOS process. However, it raises the task of transmitting multiple high-speed digital data streams from the front-end IC to the channel IC. Unless precautions are taken in the transceiver design, EMI generated by the high-speed parallel interface may significantly degrade the signal-to-noise ratio at the read head, where the desired signal is the weakest. To reduce the EMI to the read head, we propose to use multiple parallel data channels operating at reduced data rates and reduced voltage swings to carry the sampled data out of the head-disk assembly. However, crosstalk between these data channels now presents serious design concerns. Therefore, a novel crosstalk cancellation algorithm has been developed based on a compact data coding scheme by combining 8-level pulse amplitude modulation (8-PAM) signaling [3] with a data look-ahead technique. Simulation results show a 37% reduction in crosstalk among six channels each carrying 10-Gbps data stream. The design and simulation results of the key circuits in the 1-V, 0.13-μm CMOS transceiver including the samplers in the analog-to-digital converter (ADC), the phase interpolators in the clock and data recovery, and the digital-to-analog converter (DAC) output driver in the transmitter will be presented.

For high-speed digital communications, the inter-symbol interference (ISI) due to limited channel bandwidth is one of the most important factors affecting the achievable date rate [4-5]. To reduce ISI for better overall system performance, equalization in the form of transmit pre-emphasis/de-emphasis has been widely used owing to its design simplicity. However, this approach suffers from large power consumption and excessive high-frequency emission as data rate increases. An alternative is to implement adaptive equalization at the receiver front-end. Digital adaptive equalizers have been used mainly for relatively low-speed applications due to the difficulty of designing high-speed analog-to-digital converters. Discrete-time analog finite-impulse-response equalizers can achieve low-power operation and take advantage of various digital adaptive algorithms [6-9]. However, at the receiver front-end, a cross-dependence exists between the equalizer adaptation loop and the timing recovery loop. The adaptation loop only works well if a recovered clock is available. Meanwhile, as the data rate increases, the power consumption grows drastically owing to the large number of taps needed in this type of equalizer. For low-power, high-speed applications, continuous-time analog equalizers have been studied and showed promising performance [10-13]. The common approach is to adjust the frequency response of an active high-pass filter to boost the high-frequency components in the data signal relative to the low frequency ones thereby achieving equalization. Passive filters offer the advantage of zero power consumption compared with active filters. In addition, the required *LC* component values and hence silicon area become more suitable for integration with higher operating frequency. In this paper, we exploit these inherent advantages of passive filters for the design of a 20-Gb/s continuous-time adaptive equalizer.

2. Crosstalk Cancellation Technique

The suspension interconnect between the disk head and the input of the front-end IC, as shown in Fig. 1, can act as an antenna to the noise due to EMI from the high-speed parallel interface. The same noise source also causes crosstalk among the parallel interconnects. One approach to reduce inter-symbol interference (ISI) due to crosstalk is to add a compensated signal to the two nearest interfered interconnect during data transition [14]. However, the disadvantage of this approach is the sensitivity to interconnect parameter, process, and temperature variations.

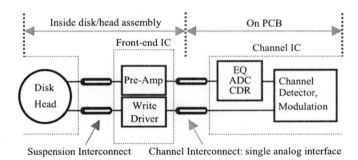

(a) Conventional architecture with analog channel interface.

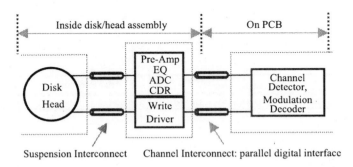

(b) Proposed architecture with digital channel interface.

Fig. 1. Block diagram of hard disk drive systems.

This work presents a crosstalk cancellation technique which employs a novel 8-4-PAM signaling with alternating 3-bit and 2-bit symbols. As depicted in Fig. 2, for the 2-bit symbols, only logic level 2 to 5 are used for data while level 0, 1, 6, and 7 are reserved for coding purposes. While the effective data rate is decreased by ~17%, 2.5-bit/symbol, compared to standard 8-PAM, the maximum data transition is reduced from 7-LSB to 5-LSB. Consequently, the EMI and thus crosstalk to neighboring interconnects are reduced. The four possible 5-LSB data transitions are illustrated in Fig. 2. The maximum transition in the 8-4-PAM can be further reduced to 4-LSB by applying a data look-ahead technique. This technique examines the next two data symbols and encodes their logic levels depending on the current data symbol. The overall data encoding algorithm is summarized in Table I. Figure 3(a) shows a specific example for the 8-4-PAM with data look-ahead technique. When Data(n)=7, Data(n+1)=2, and Data(n+2)=2, then Data(n+1) is encoded to 6 and Data(n+2) remains unchanged. As a result, the 5-LSB transition between Data(n) and Data(n+1) is avoided. A second example is illustrated in Fig. 3(b), where Data(n)=7, Data(n+1)=2, and Data(n+2)=0, the encoded Data(n+1) and Data(n+2) are 7 and 3, respectively. To decode the data in the receiver, the logic algorithm in Table I is reversed.

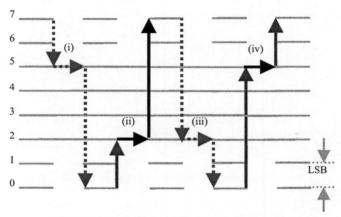

Fig. 2. Maximum data transition of 8-4-PAM signaling.

TABLE I. 8-4-PAM with Data Look-Ahead Encoding Algorithm

Actual Data			After Encoding	
Data(n)	Data(n+1)	Data(n+2)	Data(n+1)	Data(n+2)
7	2	>= 2	6	No change
7	2	1	7	4
7	2	0	7	3
>=6	5	0	7	5
<=5	5	0	1	0
>=2	2	7	6	7
<=1	2	7	0	2
0	5	6	0	3
0	5	7	0	4

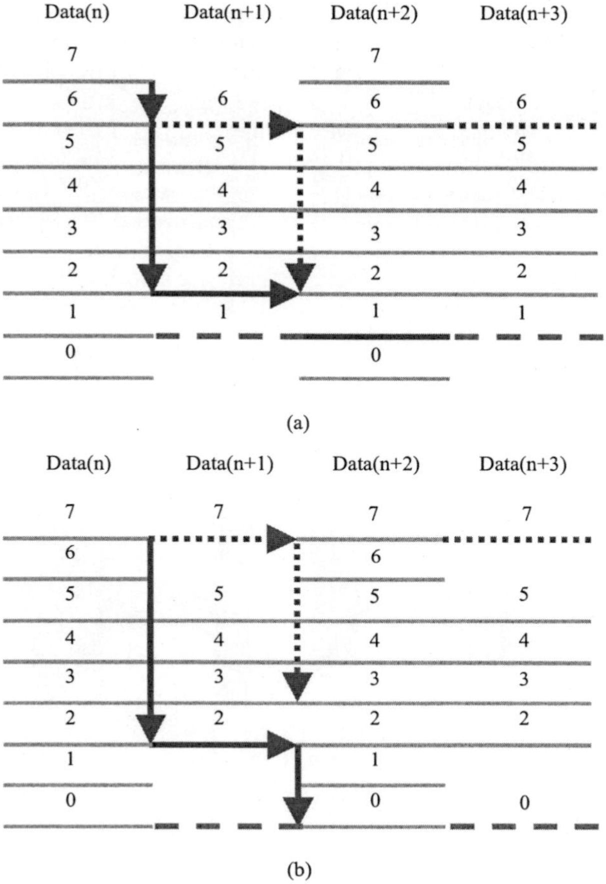

(a)

(b)

Fig. 3. Encoding examples for the proposed 8-4-PAM with a data look-ahead technique.

The effectiveness of our proposed crosstalk cancellation technique is studied by applying it to six parallel data channels each carrying 10-Gbps pseudo random data over a standard disk drive flex cable. Each 8-PAM symbol has a duration of 250 ps and a full voltage scale of 0.7 V with 0.1-V LSB. Figure 4 compares the output data eye diagrams between the 8-4-PAM with data look-ahead versus the standard 8-PAM. Our technique improves the average eye opening from 140 ps and 60 mV to 155 ps and 75 mV. A second test is performed to obtain a more direct measure of the crosstalk reduction. In this case, the third channel of the parallel interface is sending a constant logic level at 0.3-V while the other 5 channels are carrying 10-Gbps pseudo random data. The noise voltage induced on the third channel is recorded. As indicated in Fig. 5, the noise voltage observed for the case of 8-4-PAM plus data look-ahead is 0.12 Vpp whereas the standard 8-PAM is 0.19 Vpp, which translates to a 37 % reduction. Theoretically, the maximum crosstlk reduction of the proposed encoding technique is 43% ((7–4)-LSB/7-LSB). A key advantage of our crosstalk cancellation scheme is its ease of implementation in digital domain.

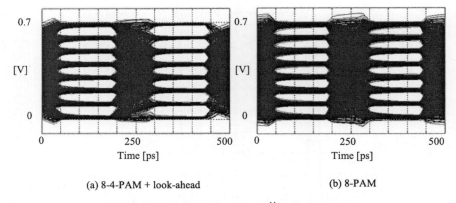

(a) 8-4-PAM + look-ahead

(b) 8-PAM

Fig. 4. Simulated eye diagram with 2^{14} random binary data.

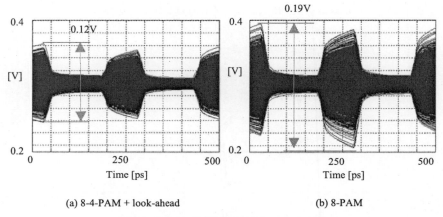

(a) 8-4-PAM + look-ahead

(b) 8-PAM

Fig. 5. Simulated noise voltage due to crosstalk.

3. Transceiver Design

The block diagram of the 10-Gbps CMOS transceiver test chip in loop-back mode is shown in Fig. 6. To support 8-PAM signaling, a 3-bit DAC and a 3-bit ADC are required in transmitter and receiver, respectively. The clock and data recovery (CDR) design is based on a phase interpolator.

The channel interconnect is modeled using the W-element in HSPICE. Figure 7 shows the data eye diagram seen at the receiver inputs during loop-back test from the transmitter through 5-cm of channel interconnects.

3.1. Receiver

The maximum clock frequency of our system is 2 GHz, or 4 GS/s. The system raw data rate is 12 Gbps with a symbol period of 250 ps using 8-PAM signaling. Two 3-bit, 2-GS/s ADCs are employed using time-interleaving techniques. The effective data rate is 10.5 Gbps after a 14b/16b encoding scheme for data redundancy.

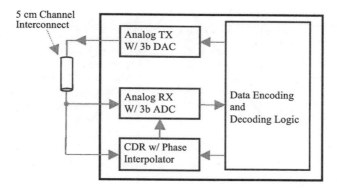

Fig. 6. Transceiver test chip in loop-back mode.

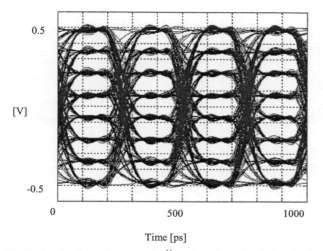

Time [ps]

Fig. 7. Simulated eye diagram with 2^{11} random binary data in loop-back mode.

The 2-GS/s, 8-PAM sampler is a critical block in the receiver. At 2 GS/s, the flat zone of each symbol is only about half of the symbol period (~125 ps) due to channel loss (Fig. 7). In general, as the data rate and the number of PAM level increase, the error probability of the integrator output becomes greater if standard integrator design is used. To resolve this issue, our receive sampler only integrates the input data during the center half of a symbol period, as illustrated in Fig. 8. The simplified schematic of the "quasi-integrator", which integrates the input signals when both Φ_1 and Φ_2 are low, is shown in Fig. 10. The Φ_1 and Φ_2 clock signals from the CDR are 90° out of phase as shown in inset of Fig. 9. The PMOS switches M_1, M_2, and M_3 are on during reset to keep the integrator outputs at V_{DD}. The quasi-integrator is followed by a conventional high-speed latched comparator in the receiver front end.

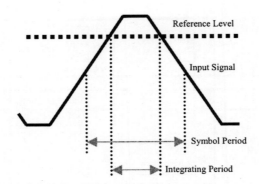

Fig. 8.   Sampling time of the integrator.

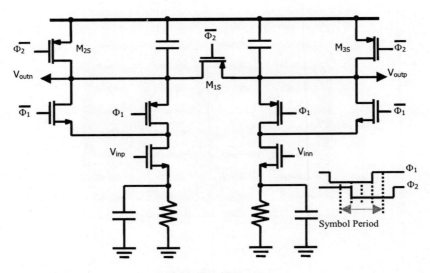

Fig. 9.   Quasi-integrator.

3.2.   *Clock and Data Recovery*

Due to its design simplicity, a phase interpolator based CDR has the advantages of lower power consumption and smaller chip area compared with a PLL based design [15]. Since CDR needs to adjust phases of the output clock signals accurately, the phase interpolator must possess good phase linearity. However, conventional phase interpolator design exhibits phase non-linearity at high frequencies owing to the variable output capacitance [16]. Figure 10 shows the simplified schematic of an improved phase interpolator. Two extra differential pairs have been added to keep the DC current levels through the resistive loads independent of the input signals. As a result, the output junction capacitance becomes insensitive to the input. The phase of the interpolator output is plotted against codes, which are digital inputs of the phase interpolator in Fig. 11.

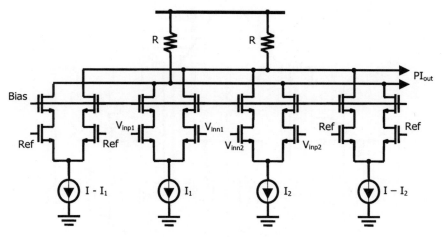

Fig. 10. Phase interpolator.

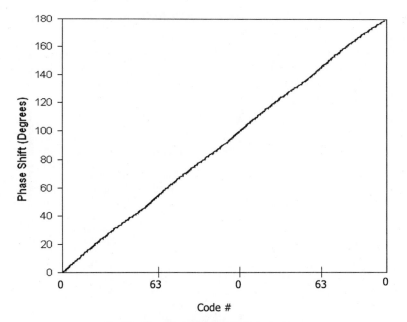

Fig. 11. Characteristic of the phase interpolator.

3.3. *Transmitter*

The transmitter output stage is implemented as a 3-bit DAC as shown in Fig. 12. The tail transistors, $M_1 - M_3$, can be momentarily out of saturation region during the crossover of the input switching. Pre-driver stages are added to steer the current smoothly from one output to the other and hence reduce overshoot or undershoot of the output signals [17]. Another approach, a data-look-ahead technique, applies to the output driver for a power-

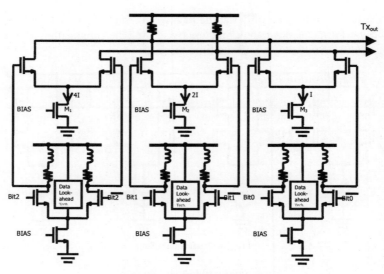

Fig. 12.   Output stage of the transmitter.

efficient driver [18], which is not associated with the proposed crosstalk cancellation technique. However, this data look-ahead technique is sensitive to process variation due to inverter delay variation. To take advantages of both techniques, our transmitter has two stages with the data-look-ahead technique applied to the pre-drivers. The transmitter output stage consumes 42 mW under a 1-V supply.

4.   Receive Front-End Equalizer Design

4.1.   *Continuous-Time Equalizer System Architecture*

To improve the receiver's immunity to ISI due to channel loss at high frequencies, adaptive equalizers can be inserted at the receiver front-end. Figure 13 shows the overall architecture of the proposed equalizer system. An adjustable passive equalization filter is

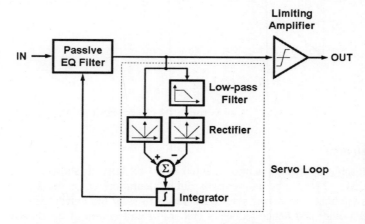

Fig. 13.   System-level block diagram of the adaptive passive equalizer.

inserted before the limiting amplifier at the receiver front-end. The feedback control loop (servo loop) performs the adaptation function by sensing the equalization filter output and providing a control voltage to adjust the filter response.

4.2. *Passive Equalization Filter*

The equalization filter is based on a high-pass *RLC* filter. The filter schematic along with the design equations and the ideal frequency response are shown in Fig. 14. This filter topology is chosen for two reasons. First, the ideal input and output impedance can be set equal to Z_0 (50Ω) over all frequencies for broadband matching. Second, even using lossy on-chip *LC* components, the filter can still function properly as long as the quality factors of all the inductors and capacitors satisfy the following condition above f_0 [19]:

$$Q > 10^{\alpha_0/40} \tag{1}$$

where α_0 is the low-frequency attenuation factor (in dB) and f_0 is the center frequency as shown in Fig. 14. Our design targets for α_0 and f_0 are 20 dB and 3 GHz, respectively. This translates to Q of 3.2 or greater at above 3 GHz. Another criterion is that the self-resonance frequency (*srf*) of the *LC* components must be higher than the upper corner frequency of the high-pass filter. This requirement limits the layout area and hence the amount of parasitic capacitance and inductance for each *LC* component. In this design, the layout of the *LC* components is optimized such that their *srf* are above 30 GHz.

A major drawback of passive equalization filters is the lack of tunability in their frequency response. This in turn makes the equalizer performance sensitive to process and temperature variations. In practice, manual adjustment is required to fine tune the filter response for different channel characteristics and operating conditions [20]. To overcome this problem, we propose to replace R_2 (in Fig. 14) by a MOSFET operating in the linear region. By varying the channel resistance of the MOSFET, the filter attenuation can be adjusted. A servo loop is then used to produce the control voltage applied to the gate of the MOSFET.

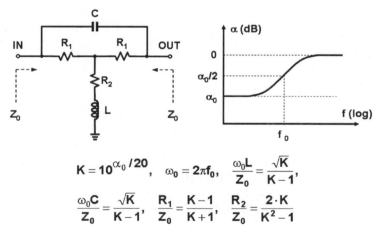

$$K = 10^{\alpha_0/20}, \quad \omega_0 = 2\pi f_0, \quad \frac{\omega_0 L}{Z_0} = \frac{\sqrt{K}}{K-1},$$

$$\frac{\omega_0 C}{Z_0} = \frac{\sqrt{K}}{K-1}, \quad \frac{R_1}{Z_0} = \frac{K-1}{K+1}, \quad \frac{R_2}{Z_0} = \frac{2 \cdot K}{K^2-1}$$

Fig. 14. Schematic, design equations, and frequency response of an ideal high-pass equalization filter.

4.3. *Servo Loop*

In conventional design of continuous-time servo loop, the feedback control voltage is generated based on the difference in the rise time between the input and output signals of the limiting amplifier in the receiver [11-12]. This technique assumes that the amplitudes of both signals are the same and that the limiting amplifier is used mainly to sharpen the signal rising edge [13]. This is not true for passive equalizers because the equalization filter attenuates the signal amplitude before the limiting amplifier restores it. Therefore, a modified continuous-time servo loop has been designed as shown in Fig. 13 (inside the dashed line). Since the power spectrum of a random date bit stream can be described by a $sinc^2(f)$ function, the ratio of the power level at any two frequencies is known [21]. The *power at two different frequencies* can be detected using a pair of narrow-band band-pass filters (BPFs) followed by rectifiers. However, BPFs typically require large power consumption especially at high frequencies. To resolve this issue, low-pass filters (LPFs) are used to capture the *power within two different bandwidths*. In practical implementation, as depicted in Fig. 13, only one LPF is actually needed because the path without any filtering simply allows the total signal power to pass through. This novel architecture essentially compares the power level of the original signal with the low-passed version. The power levels are detected using a pair of rectifiers. The power difference is then amplified by an integrator, which produces the control signal, V_C, to adjust the frequency response of the passive equalization filter for optimal gain compensation. The simplicity of the servo loop architecture greatly reduces the system power consumption.

5.   Circuit Implementation

5.1. *Passive Equalization Filter*

Figure 15 shows the schematic of the differential passive equalization filter along with the *RLC* component values. The quality factors of the inductors and capacitors at 10 GHz are 15 and 20, respectively. A PMOS device is used as the variable resistor for ease of biasing because the DC level of the signal is near the power supply voltage. The frequency response of the filter at three different control voltages between 0.5 to 1.5V is shown in Fig. 16 (a). The return loss (S_{11}) at the filter input is better than −11 dB up to 30 GHz (Fig. 16 (b)).

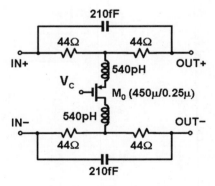

Fig. 15. Differential passive equalizer.

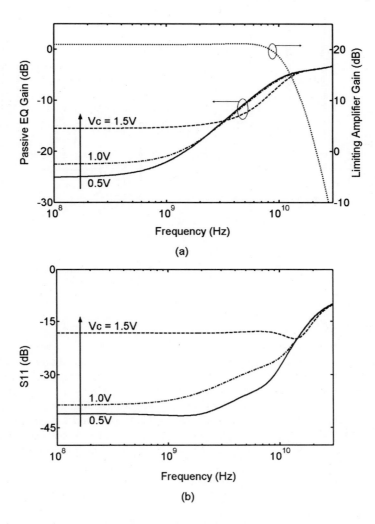

Fig. 16 (a) Frequency response of passive equalization filter and limiting amplifier.
(b) Return loss (S_{11}.) at the input of the equalization filter.

5.2. *Limiting Amplifier*

The limiting amplifier is DC coupled to the equalization filter output. It is implemented by cascading two current-mode-logic stages with shunt peaking loads (Fig. 17). The output buffer utilizes on-chip termination resistors to reduce double reflections due to impedance mismatch. The simulated gain response is plotted in Fig. 16(a). The limiting amplifier achieves a low-frequency gain of 22 dB with a bandwidth of 12 GHz. Excluding the output buffer, the amplifier dissipates 21 mW from a 2.5-V power supply.

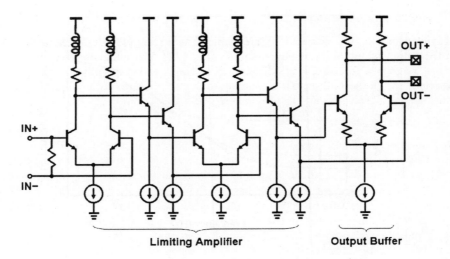

Fig. 17. The limiting amplifier and output buffer.

5.3. *Low Pass Filter*

The low-pass filter is based on a pseudo-differential amplifier as shown in Fig. 18. The active load is a parallel combination of cross-coupled pair and diode-connected PMOS devices, which alleviates the need for common-mode feedback biasing circuit and provides the necessary DC gain. M_7 operates in the linear region and acts as differential source degeneration to control the DC gain of the filter by adjusting V_G. The gain is set according to the ratio of signal powers between the all-pass and low-pass branches in the servo loop (Fig.13) for an ideal equalizer output. In practice, it is determined using a set of training data sequence. The filter bandwidth is 15 MHz and consumes 8 mW.

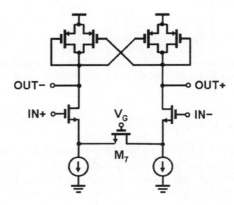

Fig. 18. The low-pass filter implemented using pseudo differential amplifier with active loads.

5.4. *Rectifier and Integrator*

As depicted in Fig. 19, the rectifiers are implemented using simple differential stages with the output taking at the common-source node [22]. The integrator has two stages. The first stage is a folded-cascode differential to single-ended amplifier to provide the necessary gain. The second stage is a common source amplifier to accommodate the large output swing. The rectifiers and integrator consume a total of 3 mW.

6. Simulation Results

The equalizer has been designed and simulated in a 0.25-μm 1P6M BiCMOS process. Figure 20 shows the transient response of the equalizer control voltage and demonstrates the ability of the servo loop to adapt for different channel conditions. The adaptation time is less than 1-μs. There are small variations in the control voltage once it has reached equilibrium, but these variations do not cause noticeable gain variation in the equalizer. Figure 21 shows the input and output eye-diagrams of the complete equalizer for both 2-m and 5-m channels. In both cases, a 20-Gb/s 2^{15} random binary data stream is used. The power consumption of the equalizer, excluding the output buffer, is 32 mW. Including the output buffer, this adaptive passive equalizer dissipates 78 mW. Table II compares the performance of our design with other recently published data equalizer results.

TABLE II. Comparison of recently reported continuous-time high-speed equalizers.

Data-Rate (Gb/s)	Power (mW)	Max. Gain[1] (dB)	Technology	Ref.
3.2	80	21	0.18-μm CMOS	[12]
5	10	9	0.13-μm CMOS	[9]
10	145	25	0.18-μm BiCMOS	[11]
20	32	20	0.25-μm BiCMOS	This work

[1]The maximum loss that can be compensated.

7. Conclusions

A new architecture using digital high-speed interface for future HDD channel ICs has been presented. To suppress the EMI problem, a crosstalk cancellation technique based on 8-4-PAM with data look-ahead encoding has been introduced. The proposed technique can be easily implemented using digital logic and is applicable to other EMI sensitive systems such as next generation compact optical modules. Designed in a 0.13-μm CMOS process, the transceiver is capable of operating at 10-Gbps while only dissipating 310 mW from a 1-V supply excluding the test logic. Furthermore, a low-power, 20-Gb/s continuous-time adaptive equalizer has been presented. The tunable passive equalization filter demonstrates significant improvements in both operating speed and power consumption when compared with other recently published results. A modified continuous-time servo loop controls the equalizer and finds the optimal channel equalization. Simulations confirm the proper operation of the equalizer for different channel conditions. Moreover, the equalizer can be implemented in a relatively small area because its *LC* components have small values and require only moderate-Q values. All of the active circuits except for the limiting amplifier have been designed using only MOS devices. The entire design could be easily ported to a CMOS technology in which the high-speed limiting amplifier is realizable.

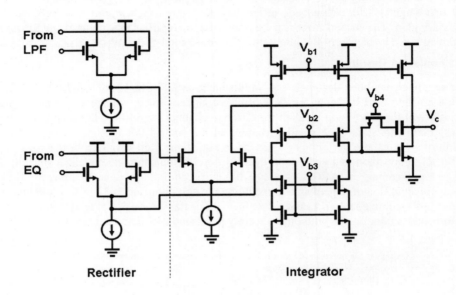

Fig. 19. The rectifier and integrator.

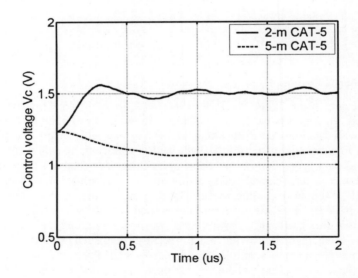

Fig. 20. Transient response of control voltage during adaptation for two different length of CAT-5 cables

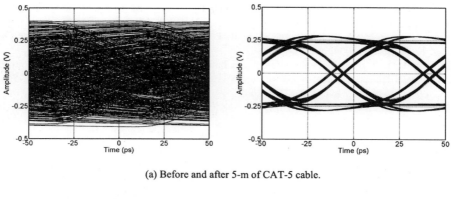

(a) Before and after 5-m of CAT-5 cable.

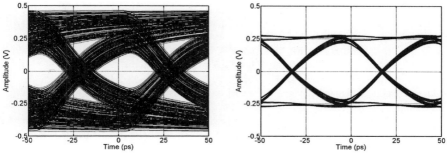

(b) Before and after 2-m of CAT-5 cable.

Fig. 21. Eyediagrams at equalizer input and output for two different length of CAT-5 cables.

Acknowledgment

The authors thank Dong Hun Shin at Carnegie Mellon University for his help in interconnect modeling. Funding support for Jaejin Park is provided by Samsung Electronics and Data Storage Systems Center at CMU. Funding support for Ruifeng Sun is in part provided by Industrial Technology Research Institute of Taiwan.

References

1. E. Grochowski *et al.*, "Technological impact of magnetic hard disk drives on storage systems," *IBM Systems Journal*, pp. 338-346, 2003.
2. J. C. Guo *et al.*, "0.13-µm Low-κ-Cu CMOS Logic-Based Technology for 2.1-Gb High Data Rate Read-Channel," *IEEE Tran. on Electron Devices*, vol. 51, no. 5, pp. 757-763, May 2004.
3. D. Foley and M. P. Flynn, "A Low-Power 8-PAM Serial Transceiver in 0.5-um Digital CMOS," *IEEE JSSC*, pp. 310-316, March 2002.
4. M. Horowitz, C.-K. K. Yang, S. Sidiropoulos, "High-Speed Electrical Signaling: Overview and Limitations," *IEEE Micro*, pp. 12–24, Jan./Feb. 1998.
5. D. A. Johns and D. Essig, "Integrated Circuits for Data Transmission Over Twisted-Pair Channels," *IEEE J. Solid-State Circuits*, vol. 32, pp. 398–406, Mar. 1997.

6. J. E. Jaussi, G. Balamurugan, D. R. Johnson, B. K. Casper, A. Martin, J. T. Kennedy, N. Shanbhag, and R. Mooney, "An 8Gb/s source-synchronous I/O link with adaptive receiver equalization, offset cancellation and clock deskew," in *IEEE Int. Solid-State Circuits Conf. Dig. Tech. Papers*, 2004, pp. 246–247.

7. H. Wu, J. A. Tierno, P. Pepeljugoski, J. Schaub, S. Gowda, J. A. Kash, and A. Hjimiri, "Integrated Transversal Equalizers in High-Speed Fiber-Optic Systems," *IEEE J. Solid-State Circuits*, vol. 38, pp. 2131–2137, Dec. 2003.

8. R. Farjad-Rad, C. -K. Yang, M. Horowitz, and T. H. Lee, "A 0.3-mm CMOS 8-Gb/s 4-PAM pre-emphasis serial link transceiver," *IEEE J. Solid-State Circuits*, vol. 35, pp. 757–764, May 2000.

9. X. Lin and J. Liu, "A CMOS Analog Continuous-Time FIR Filter for 1-Gbps Cable Equalizer," in *Proc. ISCAS*, vol. 2, pp. 296–299, 2003.

10. Y. Kudoh, M. Fukaishi, and M. Mizuno, "A 0.13-mm CMOS 5-Gb/s 10-m 28AWG Cable Transceiver With No-Feedback-Loop Continuous-Time Post Eqaulizer," *IEEE J. Solid-State Circuits*, vol. 38, pp. 741–746, May 2003.

11. M. H. Shakiba, "A 2.5Gb/s Adaptive Cable Equalizer," in *IEEE Int. Solid-State Circuits Conf. Dig. Tech. Papers*, 1999, pp. 396–397.

12. G. Zhang, P. Chaudhari, and M. M. Green, "A BiCMOS 10Gb/s Adaptive Cable Equalizer," in *IEEE Int. Solid-State Circuits Conf. Dig. Tech. Papers*, 2004, pp. 482–483.

13. J.-S. Choi, M.-S. Hwang, and D.-K. Jeong, "A 0.18-mm CMOS 3.5-Gb/s Continuous-Time Adaptive Cable Equalizer Using Enhanced Low-Frequency Gain Control Method," *IEEE J. Solid-State Circuits*, vol. 39, pp. 419–425, Mar. 2004.

14. J. L. Zerbe, et al., "A 2 Gb/s/pin 4-PAM parallel bus interface with transmit crosstalk cancellation, equalization, and integrating receivers," *ISSCC Digest of Technical Papers*, pp. 66-67, Feb. 2001.

15. F. Yang, et al., "A 1.5V 86mW/ch 8-Channel 622-3125Mb/s/ch CMOS SerDese Macrocell with Selectable Mux/Demux Ratio," *ISSCC Digest of Technical Papers*, pp. 68-69, Feb. 2002.

16. S. Sidiropoulos and M. Horowitz, "A Semidigital Dual Delay-Locked Loop," *IEEE JSSC*, vol.32, pp. 1683-1692, Nov. 1997.

17. M. M. Green et al., "OC-192 Transmitter in Standard 0.18μm CMOS," *ISSCC Digest of Technical Papers*, pp. 248-249, Feb. 2002.

18. K. Farzan and D. A. Johns, "A CMOS 10-Gb/s Power-Efficient 4-PAM Transmitter," *IEEE JSSC*, vol.39, pp. 529-532, March 2004.

19. J. T. Taylor and Q. Huang, *CRC Handbook of Electrical Filters*, New York: CRC Press, 1997, pp. 70–75.

20. "Designing a simple, small, wide-band and low-power equalizer for FR4 copper link", Technical Article HFTA-06.0 (Rev 0, 02/03), Maxim Integrated Products Inc., Sunnyvale, CA.

21. R. E. Ziemer and R. L. Peterson, *Introduction to digital communication*, 2nd ed, NJ: Prentice Hall, 2001, pp. 122–124.

22. Z. Wang, "Full-wave precision rectification that is performed in current domain and very suitable for CMOS implementation," *IEEE Trans. Circuits and Systems I*, vol. 39, pp. 456–462, June 1992.